ARCHAEOLOGY OF EAST ASIA

Animal Classification in Central China

From the late Neolithic to the early Bronze Age

AF688238

NINGNING DONG

BAR INTERNATIONAL SERIES 3031 VOLUME 4

BAR
PUBLISHING

Published in 2021 by
BAR Publishing, Oxford

BAR International Series 3031

Archaeology of East Asia, volume 4
Animal Classification in Central China

ISBN 978 1 4073 5792 8 paperback
ISBN 978 1 4073 5793 5 e-format

DOI https://doi.org/10.30861/9781407357928

A catalogue record for this book is available from the British Library

© Ningning Dong 2021

COVER IMAGE *Painted figures of twelve zodiac animals, Tang Dynasty (618-907), housed in Shaanxi History Museum, photo by Zexian Huang.*

The Author's moral rights under the 1988 UK Copyright, Designs and Patents Act are hereby expressly asserted.

All rights reserved. No part of this work may be copied, reproduced, stored, sold, distributed, scanned, saved in any form of digital format or transmitted in any form digitally, without the written permission of the Publisher.

Links to third party websites are provided by BAR Publishing in good faith and for information only. BAR Publishing disclaims any responsibility for the materials contained in any third party website referenced in this work.

BAR
PUBLISHING

BAR titles are available from:

BAR Publishing
122 Banbury Rd, Oxford, OX2 7BP, UK
EMAIL info@barpublishing.com
PHONE +44 (0)1865 310431
FAX +44 (0)1865 316916
www.barpublishing.com

ARCHAEOLOGY OF EAST ASIA

Series Editor: Anke Hein, University of Oxford, UK

This series provides a platform for data-rich studies on a variety of topics and materials from all over East Asia as well as conference proceedings reflecting the newest research insights and trends. The series takes a regional approach to East Asia, encouraging projects that cross national borders, even into adjoining regions, and/or cover areas usually overlooked in mainstream research. This includes all parts of China, Japan, Korea, Mongolia, the Russian Far East, the Tibetan Plateau as a whole and the northern reaches of Southeast Asia. Especially encouraged are submissions proposing and conducting new approaches and methods in all aspects of archaeology including scientific techniques, spatial analysis, various digital methods, but also theory and model-based or traditional chronology-focused studies.

If you are interested in publishing in the *Archaeology of East Asia* series, please contact editor@barpublishing.com.

Editorial advisory board

Ursula Brossede, University of Bonn, Germany

Katherine Brunson, Wesleyan University, USA

Lothar von Falkenhausen, UCLA, USA

Rowan Flad, Harvard University, USA

Minku Kim, Chinese University of Hong Kong, China

Hongliang Lü, Sichuan University, China

Jade d'Alpoim Guedes, University of California Davis, USA

Mark Hudson, Max Planck Institute for the Science of Human History, Jena, Germany

Xinwei Li, Academy of Social Sciences, China

Yangjin Pak, Chungnam National University, Korea

Gideon Shelach-Lavi, Hebrew University, Israel

Shinya Shoda, Nara National Research Institute for Cultural Properties, Japan

Sharon Wong, Chinese University of Hong Kong, China

Joshua Wright, University of Aberdeen, UK

A Zooarchaeological Study of the Haimenkou Site, Yunnan Province, China
Juan Wang
Oxford, BAR Publishing, 2018 — BAR International Series **2902**

The Origin of Cattle in China from the Neolithic to the Early Bronze Age
Chong Yu
Oxford, BAR Publishing, 2020 — BAR International Series **2959**

Bronze Weapons of the Qin Terracotta Warriors
Standardisation, craft specialisation and labour organisation
Xiuzhen Li
Oxford, BAR Publishing, 2020 — BAR International Series **2992**

To my parents, with love and thanks

Contents

List of Figures

List of Tables

Foreword

Taking up a position at Fudan University after my graduation, my research focus has gradually been distracted away from the subject of my PhD project. I have become busy with dealing with the faunal assemblages from many recently excavated sites. Recovered from archaeological sites of various regions and periods, they could hardly be placed under one single rubric. I experienced an entirely different way of conducting research compared with my PhD, and find myself being pushed to handle multiple tasks at the cost of the full-time devotion to one particular topic. This seems efficient and productive, in that several journal articles based on these projects can be published in a relatively short time period, while journals are always up-to-date and disseminated widely, as compared with books. Books, however, have their own advantages, for example, in unpacking complex, tedious topics requiring lengthy discussions and providing comprehensive accounts of scholarship where a focused topic can be placed in proper context. My PhD project focusing on folk taxonomy, a topic somehow rarely touched by zooarchaeologists, is exactly the type of thesis which is hard to 'dismember' and hence more suitable for publication as a book. Here it is.

This book is based on my PhD thesis, completed in 2017 at the University of Cambridge. The doctoral project was funded by the CSC International Scholarship. Additional financial support came from the Cambridge Trust, Dorothy Garrod travel grant, Newnham travel grant and Chinese Student Awards.

I owe many thanks to a great number of people and institutions who have been involved throughout the process from composing to publication. The first mention must go to my supervisor, Martin Jones, who constantly supported me with encouragement and advice during my PhD. I am greatly indebted to Preston Miracle who has donated hours of constructive discussions especially on faunal analyses. Graeme Barker and Liliana Janik provided insightful guidance for shaping the project. Roel Sterckx's suggestions pertaining to taxonomic history in China was indispensable. Peng Lü at the Institution of Archaeology CASS kindly granted me the access to his faunal datasets. Yue Li from Northwest University, China and Lingling Deng from the Chinese University of Hong Kong assisted me with their expertise in Chinese Archaeology. John Moffett facilitated access to the fabulous library collection at the Needham Research Institute. I would also like to thank all my fellow members of the Graham Clark Laboratory of Zooarchaeology at Cambridge: Minghao Lin, Janine Ochoa, Jane Sanford, Marjolein Bosch and Jessica Rippengal. I thank Pia Spry-Marques and Leah Damman for reading and commenting on an early draft of the manuscript.

Special thanks are due to Edward Allen at Fudan University for showing interest in this study and correcting my English. I also thank Ruth Fisher and Lisa Eaton from BAR for assisting me with the publication.

Lastly, I would like to express my sincerest gratitude to Jing Yuan who initiated my interest in zooarchaeology and has been the most marvelous mentor ever.

Abstract

This book explores animal categories in Central China during the late Neolithic and the early Bronze Age. It is initially motivated by the concern that imposing the modern biological classification onto faunal remains from archaeological sites runs the risk of obscuring the folk categories of the past. From this starting point, this book is devoted to addressing two issues. Firstly it attempts to develop a methodology to approach the folk taxonomies of the past by way of archaeology. By dissecting the relationship between taxonomy and deposition, it is concluded that taxonomy is a process involving physical activities and material representations, and correspondingly the archaeological approach to the issue dwells on the integration of multiple lines of evidence and their depositional contexts. Secondly this method is applied to case studies of animal classification in Central China during the late Neolithic and the early Bronze Age. Three sites on the Central Plains are involved in the study: Wadian, Wangchenggang and Xinzhai. The results demonstrate a sharp contrast between Linnaean taxonomy and folk taxonomy in ancient China. Firstly, age might have served as a critical categorical filter in ancient China. Secondly, the wild-versus-domesticated dichotomy, instead of synchronising with occurrence of domestication, was apparently only established at a date not preceding the Shang dynasty. Moreover, boundaries between the two were porous. The final trait of the animal classification in ancient China refers to its integration into a larger correlative scheme and cosmological hermeneutic. Overall this study enriches our understanding of animal categorisation in ancient China. In addition, it dissects the tension between etics and emics, language and action, domestic and wild, and finally appeals for an intellectual reflection on the connection between the present and the past as made through archaeology.

1

Introduction

Our world is filled with chaos and diversity. Arranging things into order is an important means for human beings to understand the world and their own positions within it. This process is known as 'classification'. We see classification everywhere in our life – from grocery store shelves to cyberspace.

In archaeology, typology is a common form of classification, serving as an important method to describe and explain the diversity of archaeological assemblages. While typologies are continuously questioned, given their artificial separation of material culture based on specific aims (Hill & Evans 1972; Hayden 1984), little debate, surprisingly, has centred around the classification of animal remains. Zooarchaeologists assume that Linnaean taxonomy is scientific, systematic and unambiguous (Driver 2011). The International Code of Zoological Nomenclature (ICZN) has established a universal language to standardise faunal data, which is hence compatible for cross-cultural comparisons. Such use of standardised taxonomy in archaeology, however, obscures areas of the past world where totally different classificatory systems might have been employed. Modern zoology was established in the context of the modern synthesis of biology, whose current form developed in the early twentieth century reconciling Darwinian evolution theory and Mendelian genetics. While taxonomy in faunal reports informs us on animal 'species', ancient people unlikely came across the concept of 'species' that we think of today. What they encountered were entities formulated within their own taxonomies. When we use terms such as '*Sus scrofa*' to discuss pigs in prehistory, do we in fact project our own ideas onto the ancient mind? This question is the starting point that motivates this study on late Neolithic and early Bronze Age animal categories in Central China.

1.1 Taxonomy in zooarchaeology: a reconsideration

There are two major reasons behind my research interest in prehistoric folk taxonomies.

Before this study, my major focus was on the role played by wild animals in farming societies. Despite continued exploitation of wild animal resources after the occurrence of domestication (Kent 1989; Zvelebil 1992; Hamilakis 2003), there is a surprising lack of literature addressing the issue. It would seem that such a general neglect of wild animals in the archaeological study of farming societies, especially during the Neolithic and after, has resulted from an evolutionist emphasis on domestication and a modern value judgment biased towards agricultural lifeways (Zvelebil 1992). As I unpacked this topic it became clear that the binary division of animal categories was indeed a Western ideological construct. The topic underlining wild animals and farming societies itself also fell into this evolutionist form of thinking. Therefore, before addressing the problem of wild animals, it becomes necessary to re-examine the wild/domestic division in farming societies, paying attention particular to how these animal categories were understood by these past societies, and how their divisions, if any, may have differed from our own modern constructions.

The second reason for this research centres on methodology. Faunal remains from archaeological sites are usually classified according to the Linnaean system (*e.g.* Schmid 1992: 52; O'Connor 2000: 39-40; Reitz & Wing 2008: 33-6), but the categories within this paradigm do not necessarily mirror those in the past. The role of this taxonomic approach in zooarchaeology is therefore worth reconsidering. Even nowadays, non-Linnaean folk taxonomies can be found existing simultaneous to the Linnaean system – restaurant menus, animal folk names, grouping of herds among stock farmers, to name just a few. Likewise people in the past might have categorised animals in different ways that are not reflected through the Linnaean taxonomy which, although useful, is not all-encompassing. Therefore, an alternative and complementary method has to be formulated to approach the animal taxonomies of the past. It is precisely this methodological attempt that lies at the core of this study. This methodology aims to examine an archaeological pathway towards understanding folk taxonomy.

These two concerns together lead me to the archaeological exploration of folk animal taxonomies in the past.

1.2 Anthropological approaches to taxonomy

Though Linnaean taxonomy is widely employed nowadays, several alternative classificatory systems are used in certain circumstances. While the Linnaean system is described as scientific, these alternatives are on the other hand described as 'folk taxonomies'. The study of folk taxonomy first gained academic interest among anthropologists in the 1950s during the early days of the New Archaeology movement in North America. Two mainstreams in taxonomic research – the formalist and cultural relativist approaches, gradually began to take shape – from within this movement.

1.2.1 Formalism: general principles

Conklin's study (1954) on the ethnobotany of Hanunoo culture in the Philippines initiated formalist ethnobiology.

Conklin's work recorded Hanunoo native taxa in great details and also explored their taxonomic system and economic utility.

The pioneering works in ethnobiology conducted by Berlin and colleagues (1966) have exerted a continuous and profound influence upon anthropological studies of folk taxonomy. The approach, grounded in rigorous linguistic principles, later became known as 'the school of formalists'. Formalists are convinced that the only hard fact resides in the words used to describe categories (Ellen 2006: 41) — in other words, nomenclature. A fixed name for a particular category is the outcome of the perceptual process in which the categorical conception is registered and reinforced through repeated practice and transmission among all members within a society over a considerable duration (Conklin 1962; Ellen 2006). Nomenclature is formal knowledge: a type of knowledge encoded in language, according to Ellen's definition (1999). Every single categorical group is named in the Linnaean system. It is hence appropriate to view Linnaean taxonomy as formal knowledge.

Given the importance of nomenclature, Berlin and his colleagues conducted their research primarily on the basis of categorical names in native languages collected from non-Western societies, analysing semantic meanings and hierarchical organisations in order to achieve structural regularities among ethnobiological classifications the world over (Berlin *et al.* 1973).

Berlin and colleagues summarised four general principles which are claimed to be cross-culturally applicable (Berlin *et al.* 1973). To wit: (*ibid.*):

1. Organisms are grouped according to various degrees of inclusiveness and each class (taxa, as Berlin refers to it) is assigned linguistically recognisable name.
2. Ethnobiological categories, referring to the classes which include taxa, are defined by both linguistic and taxonomic criteria. In general, they are hierarchically: 'unique beginner', 'life form', 'generic', 'specific' and 'varietal', according to Berlin's terminology. Occasionally, a sixth 'intermediate' category is found.
3. Ethnobiological categories are organised in a hierarchical system where taxa in each rank are mutually exclusive.
4. As taxonomic structure is hierarchical, taxa belonging to the same category are also placed at the same taxonomic level.

Berlin and his colleagues (1973) also drew conclusions regarding certain general tendencies shared by multiple folk taxonomies:

1. Taxa belonging to the category of 'unique beginner' are usually not labeled linguistically. This category roughly correlates to 'kingdom' category in Linnaean taxonomy in which plants and animals are not always named.
2. The ethnobiological category 'life form' which immediately includes generic taxa normally contains five to ten members. They are polytypic and labeled by primary lexemes (*e.g.* the native language counterpart of 'bird', 'fish' and 'mammal').
3. Taxa belonging to the 'generic' category are large in number, up to 500 classes.
4. Taxa in the 'specific' and 'varietal' categories are fewer than those of the 'generic' category. They are identified by only a limited number of features, if not by only one. Their names are composed of secondary lexemes.
5. An intermediate taxon is immediately included in one of the major life form taxa and includes generic taxa. These classes are also recognised as 'covert categories' since they rarely feature labeled names.

Apart from the above principles on the level of ethno-taxonomy as a whole, Berlin and his colleagues stated four principles relating to nomenclature (Berlin *et al.* 1973):

1. Generic taxa are labeled by primary lexemes. They are either terminal or immediately include taxa labeled by secondary lexemes.
2. Life-form taxa are labeled by primary lexemes. They are not terminal and immediately include taxa labeled by primary lexemes.
3. Specific names are labeled by secondary lexemes. They are terminal and are immediately included in taxa labeled by primary lexemes.
4. Varietal names are labeled by secondary lexemes. They are terminal and are immediately included in taxa labeled by secondary lexemes as well.

Berlin and his colleagues then proceeded on to demonstrate the applicability of the above nomenclatural principles in a variety of languages (*e.g.* Cantonese in China, Guarani in South America). The widespread validy of these claims persuades formalists to believe in the universal nature of taxonomies among various cultures.

In an earlier work, Berlin and colleagues (1966) also speculated on the existence of two types of taxonomic systems: the general-purpose classification where multiple organism attributes are considered and phylogenetic and genetic relations manifested, and the specific-purposed classification based on a limited set of attributes serving a a particular purpose. 'Scientific taxonomy' (*i.e.* Linnaean taxonomy) is, for example, a general-purpose system, and as such is treated as 'natural' – while a special-purpose system is perceived to be 'artificial' as it leans toward particular features of interest (*ibid.*). Paradoxically, Ellen (2008) demonstrates the inverted usage of these two terms. Scientific classification is developed by groups of specialists in the pursuit of specific knowledge while ethno-taxonomies are products of common people who establish and use them in daily life, hence being of 'general cognitive and social utility' (*ibid.*). This paradox challenges the formalist use of Linnaean taxonomy as absolute truth in comparison with ethno-taxonomies, reminding us of the particular social and historic background to the Linnaean as any other taxonomic system.

In the field of ancient Chinese ethno-taxonomic research, a highly formalist approach, modified by historiographical tradition, was adopted by Guo and her colleagues (1999) in the compiled volume *中國古代動物學史 Zhongguo Gudai Dongwuxue Shi* (*The History of Zoology in Ancient China*). Dwelling upon textual records from ancient China, nomenclature is their main source of study. In the main body of this work, Guo and her colleagues strive to match each animal category appearing in ancient writings with specific Linnaean taxonomic group(s). Consequently, such assignments, which though possibly suggestive of the degree of correspondence between the two taxonomies, failed to justify animal categories in their own context.

In short, this body of work, rigidly peformed within its linguistic paradigm, has reached some convincing conclusions, such as the universality of a concept of basic category, and pointed to cross-cultural similarities in the ways people organise their natural world. The formalist usage of native language serves an analytical device to approach ethnobiological knowledge, ananalysis of nomenclature that is basically conducted from an etic perspective. Within this framework, borders between different cultural and ecological contexts tend to be dismissed, and specific cultural configurations in varying societies are marginalised (Atran 1990). It is precisely such a weak engagement of cultural-specific variations that has resulted in this approach being contested by non-formalist anthropologists.

1.2.2 Cultural relativism: contextual variables

Resistance to the formalist school grew in the late 1970s. Scholarly opinion on folk taxonomy varied, but in general the ubiquity of social- and cultural- specific settings contributing to classification (Ellen 2006: 41; Hunn 2007) was underscored. Social constructivists such as Durkheim and Mauss, and the symbolic anthropologist Lévi-Strauss, to a certain degree, all influenced the formulation of this approach (Ellen 2006).

The work of Berlin and his associates was faulted due to its general neglect of the cultural-specific factors involved in taxonomic formation. A plethora of examples come readily to hand. The Mbalame-speaking people in Africa, for example, do not perceive the chicken as a 'bird', but instead place it under the '*chiweto*' category of domesticated animals (Morris 2000). Ellen's (2006) work on the Nuaulu folk taxonomy in Indonesia also provides an example from the cassowaries. The cassowary (*Casuarius casuarius*) is a large flightless bird. In Nuaulu taxonomy, the cassowary is placed in the same category '*manue*' as birds and bats on account of all being winged creatures, while, at the same time, also categorised under '*peni*' together with pigs and deer, considered as ritual games (Ellen 2006). Therefore, culture-specific factors are too prevalent to be treated as trivial when studying animal classifications. This sociologically framed school is known by the name of 'cultural relativism', in contrast to the 'universalism' advocated by formalists.

Studies of animal classification in ancient China do not follow either of these approaches to the full. Although studies like Guo *et al.* (1999) rely on linguistic evidence, the analysis largely follow a historiographical approach. For certain studies, the core of the research roughly correlates with formalist or relativist ideas. As noted earlier, *中國古代動物學史 Zhongguo Gudai Dongwuxue Shi* (*The History of Zoology in Ancient China*) shares simiarities with the formalist movement in its approach to ancient categories. Sterckx's *The Animal and Daemon in Early China* (2002) examines the problem from a more-or-less relativist perspective, discussing taxonomies with frequent references to their particular social background. Sterckx (2002) considers animal classification within a broad cosmos articulating different social domains, precisely underpinning the relationship between animal categories and the categories of other objects including time, space, the weather, social class and so forth (*ibid.*). In this way, animal classifications in ancient China longer belonged to some odd and remote museum collections isolated from their natural and social background but are fully comprehended within the context of ancient China where the understanding of the universe as a whole hung over the empirical knowledge of the realistic world.

In addition, cultural relativists consider taxonomy beyond nomenclature. Ellen (1999) identifies two types of knowledge responsible for the formation of animal categories. The first, formal knowledge, has been discussed in association with nomenclature above,. The second, substantive knowledge, is defined as knowledge 'which people actually apply when engaged in the regulation and extraction of resources, activities that ultimately enhance their adaptiveness' (*ibid.*). The theory of substantive knowledge ably explains the presence of the particular categorical group called 'covert category'. A covert category is a category which is not linguistically labeled (Berlin 1972; 1973; 1974; 1976). It usually occurs at unique beginner or intermediate level and can be evidently recognised by anthropologists through slip-sorting performed by informants (Berlin *et al.* 1968: 293). The slip-sorting is performed in this way: Berlin and his colleagues find that 'plant' forms an unnamed conceptual category in the Tzeltal taxonomy used among indigenous groups of southeastern Mexico. When Tzeltal informants were asked to group slips inscribed with plant and animal names based on their similarities, they could always divide animal names and plant names into two groups, indicating the existence of unnamed categorical groups correlating to 'plant' and 'animal' respectively (*ibid.*). Nomenclature therefore fails to embrace all categories. Behavioural practices, in addition, may help elucidate the entire picture of a taxonomic system. The consideration of covert categories associated with substantive knowledge again empowers relativist negotiation about contexts beyond language.

The relativistic paradigm offers no absolute truth about animal classification to which other taxonomies can universally refer. Each taxonomy (including Linnaean taxonomy) is viewed as a special cultural institution deeply embedded in particular environmental and social configuration, Linnaean taxonomy not excepted. In essence Linnaean taxonomy, as any other folk taxonomy, is likewise the outcome of a particular social and historic context. The wide employment today of Linnaean taxonomy but not Chinese or Mayan taxonomies instead, as the standard tool for identifying organisms is simply the result of the course of global history: European voyages to the rest of the world, the taxonomic interest in newly-discovered animals and plants, the scientific enlightenment of the eighteenth century, and the economic, military and ideological expansion of the West (Stearn 1973: 6–50). From this relativist point of view, Linnaean taxonomy and other folk taxonomies are self-evidently on an equal footing. Therefore, the focal point in the comprehension of a particular taxonomy lies in exploring its link with the social and historical background where it is established and utilised.

1.2.3 Tackling the tension

When different taxonomies are placed on an equal footing, each presents a method to describe the order of the world imposed by particular social and historic regulations and restrictions. We should not be surprised to find tremendous departures between taxonomies. Meanwhile similarities are shared by various classificatory schemes, as exemplified by the correspondence between Linnaean taxonomy and the Tzeltal taxonomy in Mexico suggested by Berlin and his colleagues (1966).

To explain such correspondence, Atran (1990) describes the role played by common sense in the process of classifying. Indeed Atran's theoretical evaluation of taxonomic derivations and similarities, attempts to bridge the theoretical gap between formalists and cultural relativists (Wapnish 1995).

Atran (1990: 263–4) acknowledges that the existence of a universal cognitive disposition 'determines a core of spontaneous formulated representations about the world'. Basic representations of the world are remarkably similar between cultures, it is no wonder that resemblances can also be found among taxonomies of different times and spaces, particularly in the core constituents. These universal cognitive characters therefore explain the cross-cultural similarity found among different taxonomies. In Atran's terminology, the universal cognitive disposition is exactly the 'common sense' shared by almost every human being (see Atran 1990). It is to be noted that, rather than the narrow definition specifically denoting knowledge content shared by members of a particular culture, Atran's common sense refers to the natural, basic ability of each human mind to apprehend the world and correspondingly act based on its cognitive process. Common sense processes information from the world spontaneously (*ibid.*: 264). For example, newly acquired understandings about objects themselves are conceptualised directly by common sense, including knowledge about the animals themselves. Atran (1990: 264) further suggests that a step of sophisticated conceptualisation follows the spontaneous common-sense apprehension of living things in order to elaborate information that is only partly understood, aiming at an ameliorated understanding through this process. Thoughts acquired through sophisticated conceptualisation are mainly about knowledge itself, such as ideas and notions in biology (*ibid.*). This sophisticated conceptualisation process is the starting point where diversities among taxonomies first occur.

After proposing embracing the role of common sense into the apprehension of folk taxonomy and modern biology, Atran, along with many other researchers, such as Douglas and Medin (see Medin & Atran 1999) made considerable effort to combine anthropology and psychology, aiming towards an interdisciplinary research agenda.

1.3 Previous studies on folk taxonomy: from anthropology to archaeology

While anthropological interest in folk taxonomy has increased, archaeologists have paid less attention to the subject. The result has been that only a handful of archaeological researches are devoted to the study of the folk taxonomies of the past, surely constraining our knowledge on historic and prehistoric human-and-animal relationships .

Summaries serving as general appraisals and outlooks for future archaeological work in folk taxonomy can be found in O'Connor (2013) and Sykes (2014). Both raise the issue from a larger research background and outline a series of blind points in current studies of folk taxonomies. Sykes (2014) in particular correctly points out how reliance on Linnaean taxonomy in zooarchaeology has caused us to project our own ideology onto past societies, and she urges zooarchaeologist to consider folk taxonomies when dealing with different social and cultural systems. Sykes' work also offers insightful discussions on themes of domestication, human's relationships with wild animals and ritual use of animals, calling on zooarchaeologists to reflect on larger archaeological questions.

Pioneering work integrating ethno-science, archaeology and history was conducted by Wapnish (1995), identifying folk categories in ancient Ugaritic and Akkadian texts. She argues that a concept of covert mid-level category in folk taxonomy (albeit one absent in Linnaean taxonomy) helps clarify the history of translation between the Ugaritic word *anhr* and Akkadian *nahiru*: these two words not only define one specific animal but also encapsulate the concept denoting 'part of ' or 'kind of' notions in folk categories (*ibid.*). Wapnish's research is heavily influenced by the formalist school. As a result, though she includes a section discussing faunal remains from archaeological sites, most of her arguments are essentially linguistic.

Other studies take a more or less cultural relativist attitude in tackling the issue. These arguments are firmly installed in archaeological evidence and their conclusions strongly support variation in animal taxonomy due to specific natural and cultural variations. Serjeantson's exploration of animal categories at medieval Winchester is one example. She reveals a division between food and non-food animals and within edible categories a separation between wild and domestic animals (Serjeantson 2000).

Even fewer studies touch on prehistoric taxonomy. In Marciniak's (2011) research in the Neolithic Polish lowlands, he analyses the taphonomy of faunal bones and discovers that the breaking pattern of cattle bones indicates a special manner of marrow consumption that includes roasting, breaking and cooking, while by contrast sheep/goat marrow were not roasted. He also analyses the body part representation in different animals, the result of which implies that cattle and pig were selected for certain anatomical elements whereas sheep/goat bones composed highly processed elements. The spatial pattern of bones, showing cattle and pigs were deposited in public spaces between long houses and sheep/goat bones in or around the house, again implicates two types of consumption. Linking the evidence above to ethno-taxonomy, Marciniak argues that different treatment of cattle and sheep/goat bones might correlate to a differentiation between categories of 'animals-already-domesticated' from 'animals-recently-domesticated'.

Previous studies show us the possibility to interpret faunal remains from two directions. The formalist approach pinpointing the general characteristics of the folk taxonomy is helpful in setting up a research focus and framing a theoretical structure for research, whereas archaeological data can be fully explored under the relativist approach. Therefore my interpretation of faunal remains to address folk animal categories in the past basically adopts a relativist stance, but integrates formalist knowledge if necessary.

1.4 Linnaean taxonomy and taxonomies in ancient China revisited

Having introduced different taxonomic understandings and approaches, this study intends to treat every kind of taxonomy on an equal term. No single taxonomy is more advanced or correct. Each has its own rationale rooted in a specific corresponding social and historic context. To grasp the matter fully, taxonomy has to be discussed along with the relevant background. Linnaean taxonomy and animal classification in ancient China are revisited here in their respective contexts.

1.4.1 Linnaean taxonomy

Linnaean taxonomy denotes rank-based scientific classification in general. It is credited to Carl Linnaeus (1707~1778) though its present form is not identical to the original eighteenth century concept. Carl Linnaeus organised organisms globally into a ranked hierarchy whereby the natural world was firstly divided into three taxonomic groups: the animal kingdom (*regnum animale*), the plant kingdom (*regnum vegetabile*) and the mineral kingdom (*regnum lapidum*). Kingdoms are further divided into *phyla* (or divisions for plants), which in turn are split into classes, orders, families, genera and species. The classification can be carried further, with animals arranged into subspecies and plants into *varietas* and *forma*. In the plant kingdom, Linnaeus emphasised sexual characteristics to categorise plants based on the number and the mode of the union of stamens and pistils. The theory of sexuality would be further developed and has become important to the definition of 'species' as 'groups of actually or potentially interbreeding natural population, which are productively isolated from other such groups' (Mayr 1942: xxi). Binominal nomenclature is used to name species. Linnaeus named over 6,000 species of plants and 4,000 species of animals, giving each of them a careful definition and linking them to previous literature so that the same binomial name could be matched to the same concept (Stearn 1959).

After the modern evolutionary synthesis finally took shape in the 1940s, Linnaean classification was reinterpreted against the backdrop of Darwin's theory of evolution integrated with molecular genetics. Linnaean taxonomy, however, has much deeper historic roots.

From a broad social background, the appearance of Linnaean taxonomy in the eighteenth century met the demand for an updated classificatory system needed to include the immense number of new plants and animals being discovered during explorers' voyages to exotic lands and then brought back to Europe. Aristotelian taxonomy, employed throughout the Middle Ages, failed to cover this great variety of newly discovered taxa. Plants could be examined in great detail, especially following the invention of spectacles. Aristotelian taxonomy had not concerned itself with such trivial traits. Linnaeus's system provided a consistent and concise alternative to classification and to place animals and plants in order, which consequently was widely embraced by the scientific community (Stearn 1959).

Economic needs also lay behind Linnaeus's establishment of his taxonomy. Linnaeus himself reasoned his taxonomy as economically-driven (Koerner 1996), writing in 1746, 'the task of economics is to collect [plants] from other places and cultivate such things that don't want to grow [at home] but can grow [there]' (Linnaeus, cited by Koerner 1999: 2). Linnaeus's theory of botanic acclimatisation was driven by his ambition to make Europe as rich as China in terms of plant resources by improving the harvest of exotic crops, which reflects a typical Enlightenment thought of the eighteenth century (Koerner 1996).

Linnaeus's system of classification was grounded in the assumption that God had created nature and put everything in order (Worster 1977: 39). To Linnaeus, God had created the world; his own assignment was merely to 'describe a catalogue into a methodical framework' (Browne 1983). Linnaeus was not alone in this belief, one that firmly embedded in scientists' attempt to rationalise nature since the seventeenth century, when nature was explained and explored as if it was a well-functioned universal machine under the charge of a superior power (*ibid.*: 39-41).

Additional detailed examples indicate the influence of social institutions of his time upon Linnaeus' taxonomy. Schiebinger's research (2003) demonstrates that the Linnaean classification of plants was influenced by traditional notions of gender hierarchy among eighteenth century European society. For example, the class of a plant was determined by the number of stamens, the male parts, while the order of a plant, subordinate to class, was determined by pistils, the female parts. Male parts therefore are given the priority over female parts without any empirical justification.

Overall, Linnaeus's system of classification was intimately related to the European social history of the eighteenth century, characterised by several key concepts including Enlightenment, the voyages of discovery, the development of modern science and so forth. Linnaean taxonomy was therefore an exact expression of this understanding of nature, created through the lens of then predominating social domains.

The relationship between taxonomic groups in the Linnaean system was rethought in the nineteenth century, in the wake of Darwin's publications on the theory of evolution. The principle of common descent, stating that living organisms are all related and have descended from a common ancestor with modification, further justified the 'groups-nested-within-groups' structure of Linnaean taxonomy, and further explained why this taxonomic arrangement was more appropriate than others (Queiroz 1997). The theory of evolution is primarily used to explain biological principles in the natural world. However, it always gives consideration to social, economic and political aspects. For example, natural selection was applied to social domains in the nineteenth century to justify war as a natural necessity for international selection, an excuse for the relentless warfare during that time period (Hawkins 1997: 195).

In brief, it has been illustrated that both Linnaean taxonomy and Darwinian evolution have been embedded deeply in their own particular historic contexts and shaped by social values. Despite the fact that the theory of evolution is accommodated well within Linnaean taxonomy, Queiroz (1988) views this compatibility as 'a logical consequence of the way in which taxa traditionally have been conceived' rather than reflecting objective truth about the natural world. This statement seemingly becomes even stronger when taking genetics into account, which, if necessary, can break down the basic taxonomic unit (*i.e.* species) to the molecular level. Linnaean taxonomy is widely adopted not because it is 'truer' but rather because as a consequence of the course of history and due to prevailing demands among scientific groups.

1.4.2 Taxonomies in ancient China

Ancient China appears not to have had one particular predominating taxonomic scheme. The earliest records on animal categories date back to the late Shang dynasty (ca. 1200 ~1000 BC) though information recorded on oracle inscriptions, a type of divination texts carved on bones, was fragmentary and inconsistent. As a matter of fact, in ancient Chinese texts animals were rarely discussed as major topics themselves (Sterckx 2002: 21).

A good example of this is provided by the *山海經 Shanhaijing (Classic of Mountains and Seas)*, a Chinese classical text completed around the time of the early Han Dynasty (206 BC- 220 AD). The main contents of the book are divided according to geographical categories, where information about plants and animals can also be found. The book's 18 chapters focus on specific mountains and seas. In addition to recording their geographic locations and features, the book also lists many of the spirits, animals, plants and magical objects found within these mountains and seas, among which 291 animal taxa are listed (Guo *et al.* 1999: 37). The records for these animals are attached to the descriptions of the terrestrial location, which are perceived as 'homes' for them (Dorofeeva-Lichitmann 1995). Thus the classification of the animals in the book follows their geographical spatial distribution, in turn believed to represent the symbolic scheme behind the ancient Chinese's perception of the universe rather than a realistic account of the geographic distribution of the animals (*ibid.*).

Another source of animal categories in ancient China is *爾雅 Erya (Ready Rectifier)*, the oldest Chinese encyclopedia discovered to date, comprising chapters dating from the Spring and Autumn period (771~476 BC) to the Han dynasty. The 2094 entries in the book are divided into 19 chapters. Each chapter is devoted to explaining a particular category including abstract words, kinship, architecture, utensil, musical instrument, astronomy, geography, hill, mountain, river, grass, tree, insect, aquatic creature, bird, beast and domestic animal. The final seven chapters deal with plants and animals, and are considered a valuable document of natural history in ancient China. Animals are arranged into five categories: insects, fish, bird, beast and domestic animals. Wild animals here are separeated from livestock. *Erya* indeed represents 'a kind of thesaurus or compendium of what are often cryptic glosses that were probably in origin annotations to passages in early texts' (Coblin 1993). As Sterckx (2005) puts it, 'the study of how to differentiate categories' prevails over 'zoology, the internal analysis of the categories themselves'. Rectification of nomenclature, rather than the actual classification of the animals themselves, was the primary goal of the book.

Atran (1990: 18) speculates that in early China a 'systematic attempt at taxonomic organisation' of animals was absent. This statement would only be true if the 'taxonomic organisation' solely referred to a Linnaean-like hierarchy. The organisation of each scheme is systematic and internally consistent although the classificatory criteria might have not concerned animals as biological entities themselves. For example, in the *Shanhaijing* animals are consistently arranged corresponding to the locations of

Table 1.1 Classificatory model of animals with associations of phase, season, direction and colour in ancient China (Sterckx 2002: 79)

Animal category	Phase	Season	Direction	Colour
Scaly 鱗	Wood	Spring	East	Green/blue
Feathered 羽	Fire	Summer	South	Red
Naked 蠃	Earth	Late summer	Center	Yellow
Hairy 毛	Metal	Autumn	West	White
Armored 介	Water	Winter	North	Black

their 'homes' and the classification systematically follows geography. Each taxonomy was rationalised by its specific cultural configuration.

There are actually many more taxonomies intertwined with classifications of other objects. One prominent animal classificatory system of the Han dynasty corresponds to the arrangement of 陰陽 *yinyang* and 五行 *wuxing* (the five phases) (Sterckx 2002: 72). Under this model, animals are categorised according to their covering of the skin and their categories are correlated to phase, season, direction and colour, as shown in **Table 1.1**.

On the whole, most books mentioning animal taxonomy explicitly note that animal categories are directly linked to categories of other tangible or intangible objects. It seems self-evident that natural domains were always associated with social domains. As a consequence, our understanding of animal categories in ancient China should not be isolated from their historic and social contexts.

1.5 Problems of cross-reference

As has been illustrated above, both Linnaean taxonomy and classification in ancient China above grew from their own contexts. If we take a relativist stance and perceive different taxonomic systems on an equal footing, it is inappropriate to force one taxonomy to fit into an entirely irrelevant setting where another taxonomy should have belonged. In the field of zooarchaeology then, imposing a Linnaean taxonomy onto ancient societies makes cross-reference problematic.

1.5.1 Over-identification and under-identification

The first issue concerns over- and under-identification resulting from structural asymmetry between taxonomies. As clearly illustrated by Berlin and his colleagues over the course of several publications (*e.g.* Berlin *et al.* 1973; Berlin 1976), folk categories do not always individually correspond to a Linnaean scheme. As a result, both over- and under-identifications occur when examining correspondences.

Characters from oracle bones provide one example. Oracle bones were scapula bones of ox and ovicaprids, and turtle shells prepared, inscribed, fired and interpreted for divinatory purposes in China during the late Shang dynasty. The wide range of questions are addressed to deities on oracle bones covers auguries concerning the weather, harvest, military conquer, hunting, fortunes of royal families and so forth (Keightley 1978: 33-5). Recognition of animal categories inscribed on oracle bones in the Shang dynasty provides an example of how over-identification can come about. Three characters that were found carved Shang oracle inscriptions – ,  and  – have been identified as denoting sika deer (*Cervus nippon*), muntjanc (*Muntiacus sp.*) and Pere David's deer (*Elaphurus davidianus*) respectively. The interpretation follows the stroke at the top of each character, interpreted as representing the presence or absence of, number and size of the animal's antlers in real life (Campbell 2005; Fiskesjo 2010). Whilst true that morphological differences of antlers are observed among these three species, these distinctions do not take into account the changing morphology of antler resulting from seasonal changes and animal growth. The images depicted by the three pictograms are therefore too ambiguous to point to any clear distinction, especially when the variety o oracle inscription writing styles is also taken into consideration – for instance, one character might have been written in slightly different forms by different diviner groups (*cf.* Keightley 1978).

Under-identification is also problematic. The lowest taxonomic group in the Linnaean hierarchy may not be sufficiently detailed to reflect past folk categories. Take, for example, the case of dog remains found at the Maya site of Colha om Belize. According to Linnaean taxonomy only one dog species, *Canis lupus familiaris*, was identified at the site. Their dietary consumption and spatial deposition at Colha, however, are indicative of a further division of the species beyond the definitional capabilities of the *Canis lupus familiaris* label (White *et al.* 2001).

Both examples above are the result of an indifferent application of Linnaean taxonomy to past human groups that disregards any compatability with how the animals were viewed and classified. As an analogy, we might compare taxonomy to a microscope whereas animals are the objects observed. Different taxonomies use different magnifications. As a result, one may observe one pattern of the object under one magnification and another under a different one. What our eyes are able to capture depends on the magnification under which the observation is operated. A muddle of various taxonomic magnifications runs the risk of misplacing the interpretation in a labyrinth of time, space and ideology.

1.5.2 Changeable category

Under the Linnaean system, one group of animals occupies its unique position in the hierarchical ranking system. Once an animal has occupied a position in the system, it must not be placed under another group within at the same rank. Taxonomic boundaries are thus solid within the Linnaean system.

In other classificatory systems, however, animal categories may be mutable. One object can be allocated to one or another category depending on a number of factors, including situation, participants, time, and space. Ellen (2006), for example, cites the classification of cassowary of Nuaulu people from Indonesia as evidence to support cross-classification depending on social context, given that the cassowary is placed alongside birds and bats in the category of '*manue*' on one hand, and on the other hand, it can also belongs to '*peni*', a category of ritual game that includes pig and deer. The category of the cassowary is interchangeable depending upon the social event within which this animal is discussed/exploited.

In the context of Chinese archaeology, there are also examples of categories that depended on the combination of the objects. During the Shang dynasty horses and chariots together were probably perceived as a 'compound entity' and formed an isolated category distinct from either animals or vehicles (Campbell 2015). This category was represented by their associated burials found in the great Shang settlement at Anyang and linguistically the appearance of the counter word *丙* '*bing*' (pair) implying the sense of 'paired things' (*ibid.*). Horses were considered as domestic animals during the eastern Zhou dynasty (770~255 BC), as suggested by the mention of *六畜 liu chu* (six types of livestock) in several pre-Qin documents. Although the category of horses in the Shang Dynasty has not been entirely solved, Campbell's re-depiction of the intertwining relationship between animals, material culture and social perceptions is an inspiring beginning.

1.5.3 Animals excluded from Linnaean taxonomy

The last issue concerns animals not covered by Linnaean taxonomy but classified within other systems. These animals are conventionally called 'fantastic beasts', implying that Linnaean taxonomy alone tell the truth about animals. The term 'fantastic', associated with unreal and imaginary further emphasises their absence in Linnaean taxonomy. It is the presence of such animals in other taxonomies instead that is of interest to us here. These animals may not turn out to be as 'fantastic' as their name implies if their categories are understood in association with their particular contexts and from an emic perspective.

In the case of China, dragons are the most frequently cited animals of this type. While certainly excluded from Linnaean taxonomy, it remains unexplored why they appeared in ancient taxonomic systems in China in the first place, and how people in the past perceived them. It is known that dragons were intimately linked to horses. For instance, during the Zhou dynasty, a horse taller than eight *chi* (Chinese feet) was called a 'dragon', as recorded in *周禮 Zhouli (Rites of the Zhou)*. Later in the Han dynasty, the horse was considered as the heavenly dragon's *alter ego* on Earth (Sterckx 1996). It seemed no particular distinction was made between a dragon and a horse. Both of their existences and categories could be rationalised within the particular taxonomic system which Linnaean taxonomy does not even acknowledge 'dragon'. In Medieval Europe, *Scala Naturae* includes angels and God alongside plants, animals. The Angels and God are included because medieval people believed in their existence. Are they conceptually distinct from the dragon in China? Using Linnaean taxonomy may leave no space for us to explore those animals that are excluded from the system but which occupy positions in other taxonomies.

1.6 Setting of the book: research question, region and period

The above has provided a broad introduction to the existing scholarship on folk taxonomies. As addressed above, taxonomies invariably have their roots in particular social and historical context. This opens up a venue for this current study to explore folk taxonomies in ancient China through the lens of archaeology.

The research question is largely concerned with methodology: what methods can be used to approach folk animal categories in the past through zooarchaeological data? Given the scarcity of previous studies on this subject, it is a new methodology that has to be built up. Several extant zooarchaeological methods aid the interpretation of faunal remains and each is adequate in its own way for providing sufficiently detailed information. Therefore it would be reasonable to take a conventional route sticking to these pre-existing methods, while stepping back to beg the question anew. Efforts are devoted to integrate various lines of evidence and contextualise information in order to produce a new method of interpretation.

To address this research question, a case study ideally needs to possess the following attributes. The first attribute concerns the separation of indigenous classificatory systems from Linnaean taxonomy. Separating folk taxonomy and the Linnaean system will make it easier to limit the latter as an analytical tool only. Beyond taxonomy, the separation also implies avoiding any extensive communication between the indigenous society and the Western world that roots Linnaean taxonomy. Secondly, the possibility of using information on animal categories in the local texts need not be ruled out in the case study. Though language is not used as direct evidence, it provides supplementary data which the taxonomy inferred from the archaeological assemblages may be referring to. Thirdly, instead of linguistic evidence (*i.e.* nomenclature), this project aims to explore classification in the past by examining objects for categorisation – that is, animals themselves, found in the form of rich faunal assemblages in a relatively good

Table 1.2 A brief chronology of early China, from the Neolithic to Han dynasty in the region of the Central Plains (research period highlighted in grey)

	Period	Dating (ca.)
Early Neolithic	Peiligang	7000 ~ 5000 BC
Middle Neolithic	Yangshao	5000 ~ 3000 BC
Late Neolithic	Miaodigou II	3000 ~ 2500 BC
	Late Longshan / Wangwan III Phase	2600 ~ 1900 BC
	Xinzhai	1850 ~ 1750 BC
	Erlitou	1900 ~ 1500 BC
Shang dynasty	Erligang	1600 ~ 1400 BC
	Yinxu	1400 ~ 1050 BC
Western Zhou dynasty		1045 ~ 771 BC
Eastern Zhou dynasty	Spring and Autumn period	771 ~ 476 BC
	Warring States period	476 ~ 221 BC
Qin dynasty		221 ~ 206 BC
Han dynasty	Western Han	206 BC ~ 9AD
	Eastern Han	25 ~ 220 AD

state of preservation. These three attributes allow us pin down the research region in Central China during the late Neolithic and early Bronze Age.

Unlike the situation in Europe, where the historical origin of Linnaean taxonomy can be traced back to ancient Greece and even earlier (Hopwood 1959), Chinese prehistory stands free from the development of the modern biological taxonomy. Besides, the consistency within the Chinese writing system makes it possible to recognise textual data from ancient times where folk taxonomies might have been recorded. China is therefore selected as the research region. Chronologically the research period is set between the late Neolithic and early Bronze Age. Earlier periods are not covered because the possibility of cross-reference to local textual records is ruled out in earlier prehistory. The enquiry ends in the Bronze Age given that there were increasing contacts between the east and west after this period, and so isolation of Chinese taxonomy is no longer guaranteed. **Table 1.2** briefly summarises the chronology of early China from Neolithic to the Han dynasty, focusing on the Central Plains region. The research period lies between the late Longshan period to the early Bronze Age.

More precisely the analysis focuses on the Central Plains region (**Figure 1.1**), where extensive and exhaustive archaeological excavations have been conducted and there is rich information about faunal assemblages. The Central Plains region is located on the lower reaches of the Yellow River in central China. Nowadays the region stretches out over Henan Province, south Hebei Province, south Shanxi Province and west Shandong Province. Historically, this area is believed to be the cradle of Chinese civilisation and s the political seat in most dynasties before the Song dynasty (960~1279 AD). Archaeologically, discoveries of sites spanning from the Palaeolithic to the very late Song dynasty also evidence the long cultural sequence of the region. Many important sites that witness vital changes in prehistory/history were recovered in this area, such as Jiahu site with its evidence of early animal domestication (HPIACR 1999), the Middle Neolithic village of Yangshao (Yan 1989), the Great Shang settlement of Anyang (Institute of Archaeology CASS 2007), to name just a few.

Chronologically the late Neolithic and early Bronze Age spanned from the Longshan period to the Erligang period (see **Table 1.2**), which represents a vital stage along the trajectory finally leading to the establishment of the cultural and political entity that is conventionally called 'Chinese civilisation'. As summarised in chapter five, a series of changes occurred during this period, marking an epochal departure from the preceding periods and profoundly influencing subsequent periods. These include the fortification of settlements, settlement hierarchy, intra- and inter-regional exchange, demographic migration including military excursion, emergence of a writing system and so on.

Three sites in particular, located in the Yi-Luo River basin on the Central Plain, are selected for study:e Wadian, Wangchenggang and Xinzhai. **Figure 1.1** highlights the location of the Central Plains and points out the location of the three sites in the map. These three sites chronologically fall into the same timespan – from the late Neolithic to the beginning of the Bronze Age (**Figure 1.2**).

This study examines materialised traces of taxonomy instead of linguistic evidence. The study materials hence are faunal remains — those objects themselves that were categorised. Four taxa are of main focus: pig (*Suidae*), deer (*Cervidae*), cattle (*Bos taurus*) and sheep/goat (*Ovis aries/Capra hircus*).

A series of specific questions pertaining to taxonomy in ancient China are put forward. Grasping the fact that

Figure 1.1 Map of China, showing the location of the Central Plains and the sites for study.

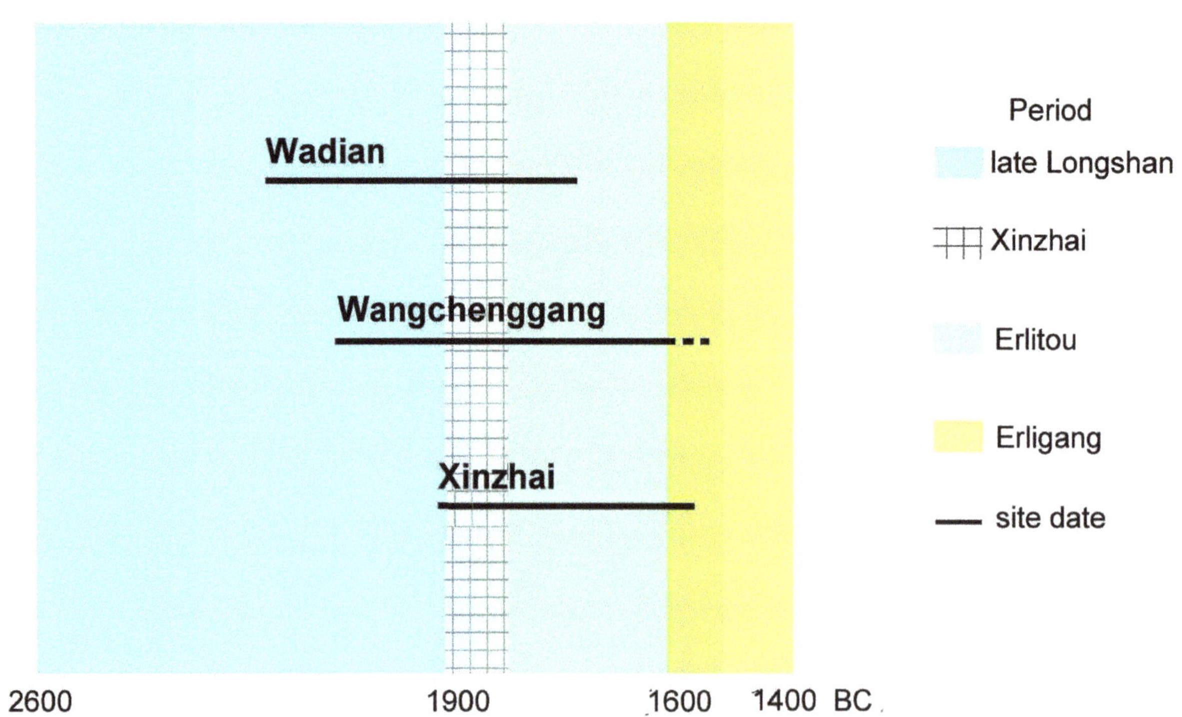

Figure 1.2 Chronology of the study sites

a reconstruction of the whole picture of a classification system in the past would be too ambitious for the current research, we approach certain more specific aspects instead. The questions under consideration include: what was/were the criterion/criteria used to categorise animals in ancient China? What was the context in which one or another taxonomy was used? By exploring these specific questions pertaining to this case study, it is hoped to provide potential solutions to the research question: how the archaeology of animal bones can inform us about folk taxonomies in the past.

1.7 Research themes

The book is organised around four themes. These key themes and their interconnections provide a vital thread to articulate taxonomy, archaeology and language and set the agenda for this research.

1.7.1 Etic and emic: two perspectives

'Etic' and 'emic' denote two contrasting perspectives to conduct research. While there are subtle differences between the definitions proposed by different anthropologists (*e.g.* Pike 1967; Goodenough 1970; Harris 1976), broadly speaking, an etic perspective conducts studies through anthropologists' – the observers' – point of view. An emic perspective, on the contrary, puts weight on internal observations and interpretations made by members of a culture and attempts to explain ideology and behaviour of a social group according to indigenous criteria (Barnard 2002). In other words, the two stances address questions from different levels: the former aims at a generalisation of universal principles while the latter focuses on justifying cultural institutions within cultural systems.

In zooarchaeology, using Linnaean taxonomy to identify faunal remains is to categorise animals from an etic perspective as the classificatory criterion — Linnaean taxonomy — is thought universally valid and appropriate to scientific research. This etic operation, on the one hand, entails cross-cultural comparisons about exploitation of animal resources whereas on the other, leaves out indigenous perceptions of human-animal relationships configured by cultural specifics. The indigenous perception, however, is the very area where an emic approach is promising.

The problem explored in this book is thus emic-oriented, that is, instead of placing animals into Linnaean taxonomy, interpretation is devoted to situating animals upon the intellectual map of past societies and rationalising the mapping within the particular context. The processing of data is still etic involving theoretical and methodological constructs intrinsic to archaeology (Berry 1989). The mutual tension is thus negotiated and it is exactly under this negotiation that this project sets forth its research inquiry and builds up its methodological framework.

1.7.2 Formalism and relativism: two approaches

The second theme dwells on formalism and relativism, the two contrasting approaches to study folk taxonomy in anthropology. Formalists, emphasising structure of a taxonomy over its contents, are devoted to eliciting universal characteristics shared by taxonomies all over the world, whereas relativists, criticising formalists' ignorance of cultural specifics, advocate that taxonomies are a kind of cultural institutions, indispensable from particular social and historic contexts (Atran 2001).

By reviewing their theoretical foundations and applications, this book clarifies the divergence manifested in opinions towards Linnaean taxonomy: for formalists who believe in a single reality, Linnaean taxonomy is deemed closer to the order of 'nature' in the world, while relativists argue that Linnaean taxonomy is no truer than any other taxonomies in that it was also constructed in a particular social and historic setting. In other words, all taxonomies are treated on equal terms from the relativist standpoint. The relativists' equivalent view of taxonomy revokes the employment of Linnaean taxonomy as the absolute truth and hence proposes the engagement of context in the study of taxonomy.

Following the thread of the discussion, this theme is to be fully developed in chapter two. By unfolding this theme, the relativist stance aims to interpret archaeological evidence anew and attempts an adequate comprehension of local categorical knowledge in the past.

1.7.3 Language and action: two pathways

Nomenclature is often employed as the proxy to study taxonomies. Both the formalist and relativist approaches mentioned above, for example, enquire into categorical names to decipher taxonomies. Categorical names are shaped by taxonomy. For example, binomial nomenclature in the Linnaean system is indicative of taxonomic hierarchy. At the same time, action is also informed by taxonomy. Looking around, one can easily find taxonomies expressed in the form of physical behaviour and material representation - shelving in the supermarket, books in the library, the layout of a zoo, *etc.* As archaeologists explore every kind of cultural practice and material product, an important hypothesis underpinning the whole of this research is hence proposed here: taxonomy informs action, and that action can be discerned in archaeological depositions. This proves the methodological potential other than through the linguistic pathway to decode taxonomies in the past.

Throughout the book, the relationship between archaeological assemblage, human behaviour and taxonomy is to be unpacked in great length. A comprehensive dissection of archaeological deposition and its relationship with behaviour and taxonomy, which I hope is clearly demonstrated in chapter three, is central to formulating an archaeological method guiding the research. This non-linguistic pathway is to be further illustrated by a number of cases studies. Finally textual/ language references, if available, are used to further test the hypothesis in addition to archaeological evidence.

1.7.4 Domesticated and wild: two categories

'Domestication' has been, and continues to be, a prominent topic in archaeology. More precisely, it is the domesticating practice rather than the category of 'domestication' that has been extensively studied. Domestication as a practice represents an important change in human lifeways. The split between hunting and husbandry, wildness and domestication, however, is envisaged within a nature/ culture dichotomy which itself is posited in the frame of Western concepts. The notion of 'wild' itself is a subject of dispute (Anderson 1997).

In zooarchaeology, a growing body of evidence regarding human-animal relationships has challenged this dualistic paradigm (Russell 2002). The division between the two may not be as rigid as conventionally conceived. Given the ever-changing definition of 'domestic animals' as manifested in, for example, medieval categories determined by animals' economic contribution to human society (Thomas 1983: 20) and Darwinian 'domestication' denoting selective breeding as an analogy to natural selection, my own conjecture is that a category of 'domestic' is created with reference to social and historic contexts and the dualism stereotype prevents a comprehension of indigenous perceptions of animal categories. Domestication of animals indicates a method to utilise animal resources. It does not necessarily have had to co-occur with a radical change in the way people viewed and ordered the world. Consequently, it runs the risk of equalling the identification of domestic animals according to Linnaean taxonomy with the creation of 'domestic' categories in ancient minds. As a matter of fact, the possibility that the wild/domestic division was absent in certain prehistoric groups cannot be easily dismissed.

Taking a relativist stance in this book, the wild/domestic dichotomy embedded in Western epistemology is to be deconstructed and reconstruction of local forms of 'domestic' and 'wild' categories is attempted by interpreting archaeological assemblages with their context *in situ*.

1.8 Structure of the book

This book is comprised of eight chapters. The present chapter, chapter one, explains the motivation behind this research and introduces both anthropological and archaeological approaches. By critically reviewing the existing scholarship, the brief history of Linnaean taxonomy and animal classifications in ancient China are revisited, rendering examples to argue for an intimate relationship between taxonomies and their particular cultural context. The recognition of this intertwining relationship alerts us to the risk of cross-referencing taxonomies and thus urges an exploration of animal classifications from an emic perspective. The research design is also presented in the introductory chapter.

Chapter two brings methodological issues into the discourse, attempting to bridge the gap between archaeological assemblage and taxonomy. Illustrated by a hypothesised example about the burial of a pet cat burial, the discussion thoroughly dissects the relationship between taxonomy and deposition, institution and agency. Intentionality appears salient to identify taxonomic depositions. The twin burial of a human and a lamb found at Çatalhöyük follows, serving as the archaeological example to be interpreted with the concern of animal categorisation. Two key concepts — contextual archaeology and structured deposition — are critically reviewed and refined to make them useful tools to decode folk categories. These two notions together guide the brick-by-brick building-up of the precise analytical methods addressed. In addition, a brief summary of the faunal assemblages under study is also found towards the end of this chapter.

More information pertaining to the sites and the assemblages under study is described in chapter three, which introduces the archaeological background of the Central Plain during the late Neolithic and early Bronze Age. On one hand, in this chapter I venture to present a synthesis of archaeological discoveries and interpretations of the study period and region. On the other hand, I explore the making of these interpretations associated with the variation in methodology and matching academic atmosphere. In this manner, the existing scholarship of archaeological studies of the research region and period is critically reviewed.

Chapters four, five and six present three case studies, each comprising an introduction of the site, a section on the result and a discussion. These interpretations are brought together and expanded in chapter seven. Three issues are discussed: classification of 'human', age-related classification and wild-versus-domesticated division. The discussion moves on to a broader scale integrating multiple lines of evidence (*e.g.* ancient texts) in order to comprehend animal categories against a wider temporal and perceptual context. For comparative studies, animal classification beyond China is briefly included.

In chapter eight, conclusions and future works are summarised. The monograph ends with a short discussion regarding its contribution to archaeology and beyond.

2

Developing an Archaeological Approach: Materials and Methods

Zooarchaeologists come across many classificatory systems. Linnaean taxonomy, employed by zooarchaeologists to identify faunal remains, is merely one of the many. As discussed in the previous chapter, Linnaean taxonomy may be inadequate to decode the folk classifications of the past. In finding a way to approach past taxonomies , archaeology can prove its potential. The aim of this chapter is to unfold the relationship between taxonomy, action and archaeological deposition, hence laying the methodological foundation of this study.

2.1 Animal categories and archaeological depositions

Taxonomy is both mental and physical. Mental, first and foremost, in that classification is an artificial system not found in natures. One may argue that classificatory principles are inferred from the natural world, but classification itself is an anthropogenic construction even if so derived. Formalists more or less treat nomenclature as the only reality in taxonomy, believing taxonomies are mental constructs. As summarised in chapter two, formalists advocate for the universality of taxonomic principles shared by human minds across cultures. The formalist statement of nomenclatural reality is founded upon the presumption of an intimate relationship between language and thought.

At the same time, taxonomy is physical. In the case of biological taxonomy, the main components of categories are both living and extinct organisms. Treatment of categories corresponds to the means of classification. Classification itself involves a process of delivering the category in the form of tangible and observable material representations. This is why we can actually witness and handle such classifications.

Precisely because the taxonomic process involves physical action and material representation, categories in the past become archaeologically approachable. In order to reveal the relationship between archaeological deposition and taxonomy, let us begin with two examples.

2.1.1 Two examples: taxonomic intention and depositional practice

An example from daily life will be illustrative at the outset. Let us presume that your neighbour's pet cat has unfortunately died. The little girl next door has buried the pet cat in the garden along with its toys and planted a flower on top of the burial. She has not yet had much experience of death in her life, except for a couple of months ago, when she went to her grandmother's funeral. Based on her experiences about death, her action for her deceased cat was the same for her grandmother.

Examined up close, this case can be explained from three aspects. The first concerns external constraints from the material world: landscape, climate and so on. In this case, the selection of the location for the cat's burial is an external constraint because the girl wanted to mimic her grandmother's funeral proceedings but could not, owing to the lack of a pet cemetery nearby. Consequently the girl had to make an alternative by burying her cat in the backyard of her house. The second aspect is related to structure or institution – that is, the social and cultural settings to which particular humans are attached. Funeral rites such as burying the corpse along with personal belongings (in this case, cat toys) as funeral goods and placing flowers above the burial can be seen as elements relating to the structure of this particular culture. All these are part of a standard funeral rite commonly practiced in the society where the little girl lives. Structure shapes the way in which individuals think and behave. As a direct consequence, the girl practiced the same funeral rite for her pet cat. Such a funeral tradition is, however, originally reserved for human beings. Her application of this human funeral tradition to the burial of her cat reflects the agency underpinning individual creativity. Burying the pet cat is an individual creation since the decision was made based on the girl's personal relationship with this particular cat as her pet, as opposed to any other cats. The three interrelated factors together gave rise to the burial of the pet cat.

When interpreting this case with a focus on the taxonomy, three categories can be identified: the human being, the cat in the biological sense, and this particular pet cat, presumably perceived as a member of the family by the girl and her close family. The first two categories can be identified according to universal biological standards while the latter can only be interpreted from an emic point of view. The category of human being provides the protocol for constructing burials while the category of pet underscores the human-animal relationship. This two considerations together guide the girl how to treat the death of her cat. The deposition of the cat described above is the result. Going further, one can also question the classifier facilitating the distinction and similarity between cat and human, the two taxa differing from each other in physical appearance and physiological behaviour, but also be grouped into the same category, presumably denoting family member? Mutual relationships rather than biological similarity likely determines the category.

Now let us revert to an archaeological example of a double burial of a human and a lamb found at the Neolithic site of Çatalhöyük (Russell & During 2006) (**Figure 2.1**). This was the only instance at Çatalhöyük where an animal was found buried under the house floor, a place where human

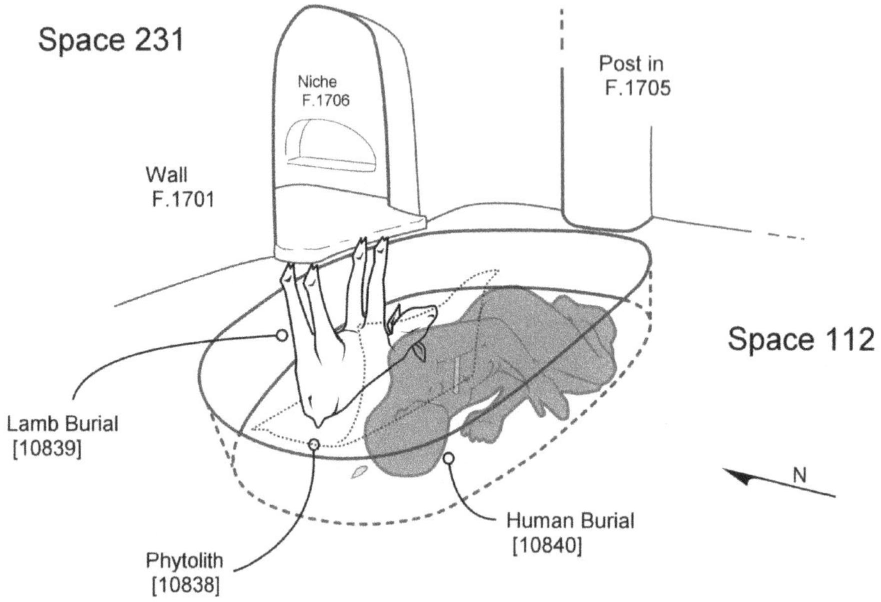

Figure 2.1 Reconstruction of the double burial at Çatalhöyük (Russell & During 2006: figure 6, John Gordon Swogger, courtesy of the Çatalhöyük Research Project)

burials are usually located. In more precise terms, the lamb had been buried next to a human skeleton with both bodies separated by a mat screen (*ibid.*). As described in the report (*ibid.*), the burial pit was dug into the house floor and measured about 1.8 m by 1.2 m. It contained a complete skeleton of an adult male, placed in at the bottom in the south. Lying on his right side, the human skeleton was placed head forwards with west and his legs flexed. Funeral goods included a bird bone tube, a large flint object and a bone point. A brown matrix covered the body and slanted downwards to the north part of the pit where a lamb was placed upon it. The matrix could be the residue of a mat used to separate the human and the lamb. The lamb was laid in a position where its legs extended vertically and its head inclined the east, suggesting that its legs were tied up as the pit was filled. Taphonomic analysis had confirmed that the lamb was fully fleshed when buried. The lamb was tentatively aged to one year old but could not reliably be sexed due to its young age.

In this Çatalhöyük case, ecology, climate and other such factors can be listed as the external constraints. If the routine construction of a burial, including the location within a house and funeral goods, guides the funeral practices in this particular society, the agency resides in interring the lamb inside the human burial. Russell and During (2006) render several possible hypotheses to interpret the human-lamb relationship in which this particular lamb might have been a sacrifice: a pet, a shaman's animal relatives, or a herd leader to symbolise the occupation of this particular male. Whatever the role of the animal, all the relationships proposed here underscore a personal bond between this particular individual and this particular lamb. This strong personal bond might have driven the inclusion of the lamb in the human burial, which apparently was not a common funeral arrangement for non-human animals at this particular place and time. The category based on this human-animal relationship unnecessarily eradicates the lamb's other category such as a sheep or, more broadly, an animal. Cross-classification is often conducted on one object. What can be asserted in this case is that the category established on the basis of personal bond was given priority over others in the performance of funeral practice and the funeral practice in turn reinforced this particular category.

A similar scenario can be seen in contemporary society where food waste containing animal bones are always deposited along with other domestic rubbish. Dead vermin like rats and foxes are usually buried or burnt and are definitely not thrown into kitchen bins. Vermin is an exact category defined according to human-animal relationships.

Thomas (2012) notes that 'all depositional practices are guided by taxonomy'. As I demonstrate in the two examples above, depositions are not only guided by taxonomy, but are actually taxonomies in themselves. The arrangement of the deposition is exactly the representation of taxonomy. Taxonomy goes beyond intangible conceptions. The direct consequence of the taxonomic process is material representation. If we imagine classifying books in a library, we not only organise these books in one's mind, but also place books on shelves accordingly. The library itself is the taxonomy. The same can be said of the pet cat example and the Çatalhöyük double burial above. The two examples evidently elucidate that depositional practice comprises one important step in the taxonomic process. Deposition is a materialised reflection of taxonomy. Thus we come to the hypothesis of this book: that deposition is informed by taxonomy. The Çatalhöyük burial in particular proves the potential to interpret archaeological assemblages from the perspective of folk taxonomy.

2.1.2 Structured deposition: definition

As the two examples above demonstrate, deposition is vital to decipher taxonomy. 'Structured deposition' represents a key concept in the archaeological discussion of deposits. Since Richards and Thomas first coined the term in 1984 (Richard & Thomas 1984), 'structured deposition' has become an umbrella concept that, though broadly employed in a variety of research, continues to lack a clear and uniform definition. It is therefore useful at this point to attempt to 'unpack' this concept so as to cast light upon taxonomies in the past, and illustrate how 'structured deposition' relates to them.

Since the early 1980s, the concept of 'structured deposition' concept has been widely used in the archaeology of various regions (*e.g.* Chapman 2000; Pearce 2008) and periods (*e.g.* Hill 1995; Clarke 1997). Gradually, during the long course of its application, its definition has grown into a palimpsest with many variations, each suiting the needs of particular sites and assemblages.

The concept was influenced by two theoretical and practical strands in the archaeology of the early 1980s: postprocessual views on the relationship of material culture and ideology, and the interest in 'ritual' among British prehistorians (Garrow 2012). In the initial stage of its application, 'structured deposition' was employed by Richards and Thomas (1984) to highlight its association with ritual activities in their study of the Late Neolithic henges of Wessex. Describing formalised, repetitive activities as 'rituals', Richard and Thomas believe that particular forms of deposition that display high levels of structure can be anticipated as signals of ritual activities, distinct from other 'daily' activities (*ibid.*: 191-2).

The association between 'structured deposition' and ritual is, however, problematic. 'Ritual' in itself is a controversial term whose meaning is open to debate. One of the most frequent misunderstandings in archaeology is accommodating 'ritual' as a 'concept' standing opposite to daily activities. This opposition is sometimes inferred by the frequency of a particular type of an archaeological assemblage in an archaeological site. Frequency alone is not persuasive enough to explain 'ritual'. In recent years, 'divergent modes of religiosity' theory (DMR theory) has challenged the stereotypical understanding of 'ritual' as equaling 'rarely occurring odd activities'. Cognitive anthropologists recognise two distinct models of ritual and religious behaviour to explain the divergence in ritual experience pertaining to memory systems: imagistic ritual and doctrinal ritual (Whitehouse 1995, 2000, 2004). While imagistic ritual involving ordeals such as violence and cannibalism occurs rather infrequently (Atkinson & Whitehouse 2011; Reinhaart 2015), doctrinal ritual, by contrast, is comprised of repetitive and routinised acts which may happen daily, including singing in choruses, kneeling during praying and so forth (Whitehouse 2002: 299). If DMR theory is accepted, ritual can be routinised as well, as is the case of Salat, the five daily Muslim prayers, a quintessentially 'daily' practice. This problem was also recognised by Richards and Thomas, who, towards the end of their 1984 paper, noted that 'domestic activity may also involve a high degree of structure' (Richards & Thomas 1984). This idea, however, is not further developed in the same paper.

In the three decades following Richards's and Thomas's publication, the concept of 'structured deposition' has undergone some profound reconsiderations. Many alternative terms to the concept of 'structured deposition' have emerged since, including 'formal deposit' (Pollard 1995), 'odd deposit' (Garrow 2012), and 'non-average deposits' (Hill 1995: 34). Following the inception of the concept by Richards and Thomas, Hill (1995) additionally unpacks the concept of ritual by listing detailed criteria to identify particular types of deposits under the banner of 'structured deposition'. The most important contribution of the work of Hill and his contemporaries (*e.g.* Bradley 1990; Bruck 1995) is that they correctly point out that everyday life can also result in structured deposition. This conceptual re-definition has been extensively employed in the interpretation of a series of prehistoric sites, especially in Britain (*e.g.* Jones 1998; Pollard 2001; Pollard & Ruggles 2001). The interpretation of 'structured deposition' was continually pushed to the limit until early 2000s when it began to be formally criticised (*e.g.* Bruck 1999), after which we see the academic resurgence of the study of 'everyday life' in archaeology (*e.g.* Thomas 1999; Pollard 2001).

More recently, Garrow (2012) has thoroughly reviewed the development of the concept of 'structured deposition' and addressed three critical issues as a recapitulation of this 30-year debate: 1) the unpacking of various types of deposits caused by different formation processes under the glossary concept of 'structured deposition', 2) the reconsideration of the 'meaning' coded in practice and 3) the inclusion of both ritual and everyday practices represented by material culture (*ibid.*). This seemingly implies that 'structured deposition' occurs much more readily than presumed and that any direct connection to 'ritual' requires a reassessment given the rather hazy border between ritual and daily life. Therefore, 'structured deposition' is determined by neither the display of structure in deposition nor the frequency of archaeological assemblage. The two factors inform us about the patterning of archaeological assemblages in depositions. Intentionality, I propose, is the most important factor in comprehending the nature of these so-called 'structured depositions'.

2.1.3 Intentionality

Not all depositions displaying structures are 'structured depositions'. Structure as a pattern is readily found in archaeological assemblages. In addition to anthropogenic agents, geomorphological factors such as weathering and fluvial processes may also produce structures given the mechanical nature and physical laws governing their influence upon deposition formation. Emphasis upon on the structure manifested in the patterning of the assemblage somehow downplays the concept of 'structured deposition' towards a mere expression to describe the phenomenon of the patterning of archaeological assemblages. Obviously some natural mechanisms may result in a repetitive pattern within a deposition, one which also displays some variety of 'structure'.

I concur with Garrow (2012) that different types of deposition should be placed along a spectrum, retaining 'structured deposition' as a matter of scale. This proposal echoes my emphasis of 'intentionality' in the deposition formation. From this point of view, refuse depositions, though usually ranking at the lowest end of the scale, are formed intentionally.

Garbage depositions are intentionally constructed under a binary taxonomic distinction of valuable-versus-valueless. The distinction is sufficently persuasive to predominates in archaeologists' mind when deciding on excavation collection and discard. Archaeologists come across a wide range of objects during their excavations. Pottery, stone tools, copper artefacts will be gathered, rather than fluffy dust, muddy soil (although certain quantity of soil is sampled for environment-related studies) and tiny pebble fragments, being considered as useful evidence to inform us about the past. This selection of particular categories is truly a projection of our own perception, shaped by the same powerful binary classifications of the past, leading us to determine archaeologically valuable from valueless. Under the valuable-versus-valueless binary, two general categories emerge, broadly separating refuse dumps from those depositions containing valuable items. Our definitions of refuse become themselves immaterial. All refuse, first and foremost is valueless and useless. Before one decides to dispose of an object, it must be decided first whether this particular object is rubbish. Therefore both decisions are intentional, and the construction of a rubbish dump intentional as well. In most cases, one will find that though rubbish disposal is intentional, it lacks clear structures, as no further curation of the dump is required after disposal. This may be true in most but not all cases. In some ethnographic examples structures are easily recognised in rubbish dumps, such as the Marakwet group living in modern Kenya, Africa, recorded by Moore (1986: 102), where a disposal pattern is associated with the social connotations of things (*e.g.* separation of ash, chaff and animal dung according to their age and gender connotations), one, however, is performed unconsciously out of habit (Garrow 2012; Thomas 2012).

Apart from rubbish depositions, other depositions that are intentionally constructed for purposeful utilisation are indicative of the categorical schemes behind valuable objects. In fact they unfold more detailed and complicated aspects of taxonomy as obviously people elaborate categories of valuable things that of potential use in their life. In certain circumstances, not only is the deposition of objects of which the immediate result is the structure displayed by the assemblage intentional, the patterning of the structure itself is also intentionally arranged as an indispensable part of the taxonomy.

My emphasis of the intentionality in the formation of structured depositions aims to pinpoint the nature of structured deposition. It does not provide a pragmatic guideline for the identification of structured deposition. Frankly, I remain skeptical whether any so-called 'guideline' ready to identify depositional type can exist, and ultimately whether this 'guideline' — possibly a long list of traits — can genuinely assist our understanding of the formation process of archaeological depositions. If a list of quantitative criteria alone is insufficient to interpret depositions, contextual analysis may provide a supplementary direction for further investigation.

In short, 'structured deposition', revisited in this study refers to those depositions arising from taxonomically framed intentionality. The term, subsequently, becomes widely inclusive given that all depositional practices are guided by taxonomy (Thomas 2012). Nonetheless it is exactly the difference in structure, content and other factors that can inform us about folk categories. To be clear, recognition of structured deposition is not the priority of this project. Instead, recognition of the *intention* behind the depositional action is the issue of central concern. Depositional intention only acquires meaning in its particular context. Hence an explanation dwelling on a case-by-case specification is favoured here as an echo to the archaeological interpretation based upon the combination of quantitative and contextual analysis.

2.2 Contextual archaeology

The contextual approach proposed in this study represents a modification of classical contextual archaeology methods.

Contextual archaeology was initially proposed by Ian Hodder in his book *Symbols in Action: Ethnoarchaeological Studies of Material Culture* (Hodder 1982a). In Hodder's own words, contextual archaeology concerns 'the role of material cultural in the reflexive relationship between the structure of ideas and social strategies' (Hodder 1994). Agency and meaning are two core concepts in Hodder's theoretical framework. The highlighting of agency elucidates the specification and furthers its mutual interaction with the more general course of history, and the proposal of meaning calls for the kind of archaeological interpretation that is incorporated with original past contexts. Contextual archaeology forms one part of postprocessual archaeology as opposed to processual archaeology, which in general employs science-concerned, model-based approaches to seek a universal generalisation regardless of historical and cultural variations (Barrett 1987).

The emphasis of 'context' in contextual archaeology inspires the approach developed in this book. Nonetheless it is not an exact copy of classical contextual archaeology methods. Unlike contextual archaeology, which is devoted to explaining the precise 'meaning' in the past at most, symbolic meaning of animals or animal categories is not the premier object of this project. Besides, the project discusses agency with the focus upon its dialectical power to influence the course of the wider entity such as community and society rather than its minute recording of individual lives. In other words, I retain specification as key to understand animal taxonomies site by site and attempt for an interpretation of faunal remains articulated with other domains beyond direct connections with animals, such as the use of space and intra- and inter-settlement social network, to name just a few issues.

Contextual archaeology appeals for intellectual reflection from researchers regarding their own interpretations, which benefits the methodological development of this study. It remains important for archaeologists to understand the boundary between our own epistemology and those ideas installed in the remote past. This concern, addressed in contextual archaeology, is carefully dealt with in this book.

In the pursuit of contextual archaeology, two types of analogies in archaeology put forward by Hodder (1982b) also fuel the discussion on the limits of etic Linnaean taxonomy to examine classification in the past.

The first type of analogy is termed 'formal analogy'. 'Formal analogy' is performed on the assumption that 'if two objects or situations have some common properties, they probably also have other similarities' (Hodder 1982b: 16). Zooarchaeological identification of animal bone is a type of formal analogy. Because animal bones from archaeological sites share morphological similarity with modern reference bones, they most probably belong to the same biological category. This determination of animal category follows the logic of formal analogy. Hodder (1982b: 16) criticises formal analogy as subject to 'fortuitous' or 'accidental' similarities and uses 'style' as an archaeological example to argue against its neglect of social contexts (Hodder 1990a). While methods strengthening formal analogy do exist, such as increasing the number of aspects for comparisons, Hodder (*ibid.*: 16) refers to alternatives in the analogy and relational analogy, which not only examines covariation itself, but also discuss 'nature and cause' (*ibid.*: 27). By bringing the context of covariation into consideration, the reliability of analogies is improved.

The notion of relational analogy is useful to this study. Considering the fact that animals do not undergo dramatic morphological change, formal analogies may seal the possible categorical shifts. Relational analogy associates similarities between the past and present of their contexts in which they are interdependently connected. The notion of context again resonates with the approach proposed in this project.

Hence in short, contextualisation is central to this project. Faunal remains are firstly contextualised in archaeological sites to interpret their treatment and deposition. Thereafter animal categories are contextualised in their own cultural background in order to assess their utility and influence.

2.3 Zooarchaeological materials

With the theoretical framework for this study outlined above, the datasets for study can now be introduced. As stated in chapter one, detachment from the cultural context of Linnaean taxonomy, together with the potential access to local textual records, has resulted in the book focusing on the late Neolithic and early Bronze Age Central Plains region in China. Depositional action, which directly results in the patterning of assemblages from archaeological sites, is informed by taxonomy. Faunal remains indicative of animal taxonomies therefore form the material for study.

Three major criteria were considered when assessing the suitability of faunal datasets for the current project. First and foremost, only those assemblages in a relatively well-preserved condition were selected for study. Secondly, assemblages with rich contextual information were favoured. Thirdly, assemblages that shared a certain cultural continuity with each other were selected so that interpretations of their archaeological remains could be easily and adequately compared and synthesised. Based on the three selection criteria above and the zooarchaeological collections available, it was decided that this study should focus on archaeological sites in the Central Plains of China between the late Neolithic and early Bronze Age.

Generally speaking, the environmental conditions of the Central Plains are conducive to bone preservation. The region is considered as the cradle of Chinese civilisation and has been targeted for extensive and exhaustive archaeological surveys and excavations that have yielded rich archaeological information. During the late Neolithic and Bronze Age, intensive social interactions were underway in this region, forming a social landscape of closely intertwined societies, giving rise to robust archaeological records within which the faunal data could be clearly contextualised. The sites Wadian, Wangchenggang and Xinzhai illustrate this particularly strongly. Geographically, the three sites are located along the Yi-Luo river valley on the Central Plain (**Figure 2.2**). Dates obtained from the three sites fall perfectly into the chronological sequence from the late Neolithic to the early Shang dynasty. All, more significantly, have yielded a wealth of animal bones appropriate for this study. A more detailed introduction of the sites is presented in chapter three.

In this book, we collate data from from published faunal reports (Wangchenggang site: HPIACR and MCH 1992, SAMPU and HPIACR 2007; Xinzhai site: ACSAC and ZMIAC 2008); from original recording sheets provided by associated institutions and excavators (Wadian site: Dr Peng Lü from the Institute of Archaeology, CASS), and from the Chinese Archaeology online fauna archive (part of Wangchengang faunal dataset: Chinese Archaeology, 2013). In order to build up a consistent and comprehensive database,

Figure 2.2 Map of Henan Province in Central China, showing the locations of three archaeological sites for this study

faunal reports from each site were re-edited, recording information about taxon (to species, if possible), element, element part, side. and context, and redundant information were deleted. Four major animal taxa were involved in the analysis. They are pigs (*Sus*), deer (*Cervidae*), cattle (*Bos taurus*) and sheep/goat (*Ovis aries/Capra hircus*). Depending on the state of bone preservation, other information, such as epiphyseal fusion, tooth eruption and wearing, and surface modifications, was recorded: all such information provide additional information on the treatment of animals.

2.4 Zooarchaeological methods

Research methods play a critical role in articulating the question posed and available data. Given the lack of suitable pre-existing methods for the current study, methodological refinements in which both the nature of the faunal assemblage and data availability are taken into account have been deemed necessary. The following methods or precisely pragmatic techniques are established within the methodological framework stated above.

2.4.1 Identification and recording

'Taxon', 'element', 'side', 'element portion' and 'excavation context' comprise the five basic recording units. In the original reports, 'identifiable' specimens included bones where the 'element' category was known but the taxon remained unclear. Given that the research aim of this book resides in taxonomy, 'identifiable' specimens here refers only to bones that were both anatomically (able to determine the element) *and* taxonomically (able to determine the taxon) diagnostic. Admittedly, this redefinition of what is identifiable is likely to be affected by degree of fragmentation. On the other hand, inclusion of those element-identifiable, but taxonomically-indeterminate specimens adds a degree of frustration on the part of the researcher when performing quantitative analysis. Unlike some other zooarchaeological identification systems where 'diagnostic' and 'undiagnostic' categories are established in order to house such bone types in addition to 'identifiable' specimens (*e.g.* Miracle & Pugsley 2006: 260), little information can be extracted from these taxonomically indeterminate element types under current identification protocol. Therefore the strategy here, determined by the nature of the archaeological data and guided by the research question, involves honing in on data with the finest resolution possible, so as to achieve a fuller understanding of how animals were treated corresponding to their categories, if at the expense of losing the potentially significant information residing in more stochastic data.

2.4.2 Quantification

The Number of Identified Specimens (NISP) is the most basic quantification unit employed here. Its use has been criticised many times and it is known introduce biases as a result of differences in fragmentation levels, recovery strategies, the skeletal anatomy of taxa, and so forth (White 1953; Shotwell 1958; Chaplin 1971; Grayson 1984; Lyman 2008). Despite of all these critiques, NISP remains the most commonly used quantification unit. For this study in particular, as identifiable bones are redefined, referring to those bones which are both anatomically and taxonomically diagnostic, NISP remains an explicit quantification unit to employ.

In essence, NISP is the most straightforward and rudimentary counting method from which other zooarchaeological counts and estimates, namely MNI (Minimum Number of Individuals), MNE (Minimum Number of Elements) and so forth are derived.

Information on the Minimum Number of Individuals (MNI), included in the original faunal reports, is not employed in this project. This is because, firstly, the procedure inevitably suffers from biases relating to fragmentation, as the great majority of faunal specimens from archaeological sites experience fragmentation effect to various degrees (Grayson 1984: 25; Lyman 2008: 43-4). Secondly, MNI estimations can be potentially affected by sample size: especially for smaller-sized samples, where the over-representation of less abundant taxa becomes problematic (Grayson 1981). Thirdly, MNI is also subject to the level of aggregation (Lyman 2008: 57-69). It is heavily dependent on the scale of the unit upon which MNI is being calculated: in general the finer the scale, the more MNI estimates will exist for a single site (Watson 1979; Grayson 1984). Fourthly, MNI was first introduced to estimate the volume of meat rendered by animals (White 1953) rather than as a substitute for NISP. In other words, this analytical calculation served special purposes. Last, but not least, as MNI is derived from NISP, it also suffers from all the biases noted above known to influence NISP counts. MNI is therefore not used in our study. This is not to say, however, that the idea behind the estimation of 'number of individuals' is not valid. On the contrary, it is an important concept, especially when interpreting the nature of different deposits. Therefore, when deemed necessary, data about individuals will be presented in the form of descriptive information.

2.4.3 Reconstruction of data

The taphonomic, demographic and utility data analysis presented here is performed using NISP counts.

Fragmentation

The degree of fragmentation of an assemblage can be recorded and evaluated in many ways (*e.g.* Dobney & Rielly 1988; Outram 2001). A close examination of fragmentation heavily depends on the precision of the recording methods used. Though faunal data for this study is a stronger source of information than that for other regional sites, such information remains insufficiently detailed with regard to specimen zonation, which consequently does not allow for finer-scale fragmentation analyses. This coarser resolution means that the level of fragmentation has to be calculated by taking the proportion of NISP against the Total Number of Bone Fragments (TNBF hereafter, including both identifiable and unidentifiable bones). NISP: TNBF is calculated for each context in order to unveil depositional differences between the contexts and for scaled comparisons to be performed.

Age-at-death

Age profiles can inform us on husbandry and hunting strategies. The age at death of an animal is estimated by examining tooth eruption and attrition patterns, aided by epiphyseal fusion data. Age profiles of pigs and deer were estimated.

Stages of dental development in pig mandibles were assessed using Grant's (1975; 1982) method. As information about tooth wearing is not fully recorded in the faunal reports available, dental eruption alone fails to provide data of a sufficiently-fine resolution. As a result, relatively broad age groups were established to present age profiles. In pigs, five age groups were reconstructed according to O'Connor's (1988) method: juveniles (< 5 months), immature individuals (5~10 months), sub-adults (12~24 months), adults (30~36 months) and old individuals (> 36 months).

Age estimations based on mandibular dental eruption were also conducted on the deer specimens. Deer presented in this project largely include sika deer (*Cervus nippon*) along with Pere David's deer (*Elaphurus davidianus*) and a limited number of small, unidentified deer. The specimens were divided into two age groups: those under 24 months lacking full teeth eruption and those older than 30 months with complete sets of teeth. As this division produces very broad age groups, the difference in dental development between different deer taxa is not significant under such coarse resolution. Therefore the ageing method used and the ages for dental eruption were modified from those proposed by Carden (2006) and Wang *et al.* (2014) regarding red deer (*Cervus elaphus*) and sika deer (*Cervus nippon*).

Apart from dentition, the degree of epiphyseal fusion presents another useful gauge of age at death. Studies have shown that epiphyseal fusion data is compatible with dental indicators of animal age (Zeder 2006). Compared with dental eruption and attrition, epiphyseal fusion provides somewhat coarser age estimates. This methods also suffers from problems relating to differential destruction (O'Connor 2003). The employment of epiphyseal fusion data here serves as a supplementary source to the dental age estimations, preventing over-reliance on a single ageing method. Bones were recorded as either 'unfused' or 'fused'. Following Habermehl's (1975) method, bones were divided into three groups according to their fusion ages: early fusion (0~10 months), middle fusion (18~24 months) and late fusion (>36 months).

Taxonomic composition

A faunal assemblage usually contains more than one animal taxon. The proportional representation of each of the taxa is based upon NISP counts. A general description of taxonomic abundance is presented at the site scale, followed by, where appropriate, comparisons between contexts.

Given the significantly larger number of pig and deer remains over those of other fauna at Wadian, a differentiation index between pigs and deer was established to evaluate abundance for either taxa. The index was calculated as follows:

$$INDEX = \frac{NISP_{pig} - NISP_{deer}}{NISP_{total}} \times 100\% \textit{ (in each context)}$$

Indices from contexts were divided into three groups considering the total sample size of each phase and the percentage of pig and deer remains. The first group consists of those contexts where pig remains predominate, but deer bones are totally absent (index ≥ 0.6) ; in the second group the indices stand for assemblages containing slightly more pig remains than deer; the final group includes those contexts that have more deer remains than those of pigs (index ≤ 0). Different shades of grey are used to differentiate the three groups: dark grey is used for the pig-rich group, light grey for the deer-rich group and moderate grey for the neither pig-rich nor deer-rich group in between. The groupings reflected in the different colour are then visualised on the site map for each period. This pig-deer index is not applied to other sites given the fact that assemblage compositions at other sites are likely not suitable for this method.

Body element distribution

The distribution of bone elements elucidates humans' treatment of animal carcasses. Factors such as humans' dismembering skills, transport strategies, and body part selection based on utility can all shape the patterning of bone element distributions. Here, bone elements are divided into five anatomical regions: head, axial, forelimb, hindlimb and extremity. The head region includes cranial bones and mandibles, excluding loose teeth. Vertebrae, pelvis and sacrum belong to axial elements. The forelimb comprises the scapula, humerus, ulna and radius, while the hindlimb consists of the femur, patella, tibia and fibula elements. Metapodia and phalanges constitute the extremity group, even though in some studies metapodia are grouped with limb bones (*e.g.* Stiner 1991; Pickering *et al.* 2003;). Technically speaking, metapodia are bones of the animal foot. They are here included in the extremity region based on their anatomical locations and so as to avoid any presumption regarding their utility and function. For each of the sites comparisons are made between taxa, contexts and periods. Their similarities and differences are tested statistically where appropriate. Paired Wilcoxon tests will be run on the PAST program (Hammer *et al.* 2001) and the exact p value will be used as the indicator.

Associated bone group (ABG)

An associated bone group refers to a set of skeletal elements from one individual animal deposited within one context (Morris 2010). The early version of its definition, such as put forward by Grant (1984) and Hill (1995), includes deposits of single skeletal elements, fragmented skulls and complete mandibles. Such unusual animal depositions were often interpreted in association with 'ritual' in the studies concerning Iron Age Britain though the precise meaning of 'ritual' remains ambiguous. When used here, ABG aims firstly, to distinguish clearly between different types of depositions, enabling the interpretation of the formation process. Secondly, in some cases, it works well as an alternative concept to the MNI estimate, in that it avoids the over-representation of individuals based on body parts.

2.5 Analysis of contexts

As was discussed in chapter two, the contextual approach employed here is different from classical contextual archaeology. A few methodological refinements have been made. Firstly, 'contextual' here stresses the connection between archaeological fauna and *in situ* depositional contexts. Secondly, the time-depth within a faunal assemblage is also evaluated. The pathways through which animals interacted with humans and finally ended up in archaeological deposits are viewed as an impartible part of the information carried by bone specimens. Thirdly, instead of envisaging the exact 'meaning' of animals in the past, context here is more capable of telling the consequence of the 'meaning' – that is the treatment of the animal guided by the meaning attached to the animal category (*e.g.* how animals under particular categories were used instead of why these animals belonged to particular categories). In short, the contextual approach used here advocates an interpretation of faunal assemblage associated in its original social and cultural context by examining its archaeological context.

2.5.1 Spatial analysis

Spatial analysis forms a significant part of this approach as it detects the relationship between the animal category and the larger classificatory scheme. Spatial analysis is performed on two scales. Firstly, different types of deposits are identified in order to explore the formation processes behind the faunal assemblages. Attention is paid to differentiating the treatment of animals through the depositional process. Secondly, at the site level, faunal depositions are interpreted based on their spatial location within the site.

2.5.2 Isotope data

Isotope analyses have been widely conducted to cast light on human-animal interactions (for a review see Lee-Thorp 2008). Isotopic signals are also suggestive of categorical differentiations which otherwise would not be detectable (*e.g.* White *et al.* 2001). Due to the scope of this study, no new isotope data has been obtained. Existing isotope data from the study region have been reviewed and re-interpreted paying special attention to how these might be used when establishing and analysing animal categories.

2.5.3 Lexical indication and written record

As illustrated in chapter one, one critical way to approach folk taxonomy is through linguistic analyses. In addition to the archaeological data at the core of this study, lexical information and written records are also noted in the discussions of the material when necessary. The fact that ancient characters and literatures may represent and provide a glimpse into what the animal-human world was like in the past cannot be ignored, taking Proto-Indo-European lexical indication for lifeways on the Steppe for instance (Benveniste & Lallot 1969; Jones 2002). Rather than serving as direct evidence, lexical information is treated as an additional source for reference in parallel to the archaeological observations, since we are aware of the gap between the first appearance of the Chinese writing system and the period under study here, and the palimpsest nature of ancient literatures.

2.6 Summary

To sum up, this chapter has unveiled the relationship between taxonomy and deposition firstly. Beginning with two examples, it has demonstrated that depositional practices invariably correspond to taxonomy. In fact, deposition itself is one salient part of taxonomy. As taxonomic process consists of both physical activity and material representation, the gap between animal category in the past and patterning of archaeological materials thus is bridged, which leaves substantial potential for archaeology to explore animal categories in the past.

A pair of archaeological concepts 'structured deposition' and 'context' –cast light upon the methodological development of this study. It is 'intentionality' that determines the nature of a deposition. When 'structured deposition' is considered with reference to taxonomic intention, it ends up being much more common than conventionally suggested. 'Context' is still a key concept to this book. The identification of archaeological fauna according to Linnaean taxonomy detaches from archaeological contexts. Such analogies, therefore, are fairly weak. Depositional contexts containing rich information about how animals were treated thus survive as a key factor to this research.

Having established the theoretical framework, the second half of the chapter outlines in detail the techniques employed for analysing the faunal data available. Though quantification is simplified and reconstruction of demographical data is restricted around a rather coarse resolution, the inclusion of contextual analyses aids the interpretation of the faunal remains associated with their archaeological and emic contexts. This method is applied to three archaeological sites in the research region in central China, and these are: Wadian, Wangchenggang and Xinzhai sites. Before proceeding to the results from these three sites, the review of the archaeological achievements on the late Neolithic and the early Bronze Age central China that informs the next chapter will profoundly facilitate our understanding of this research period and region, thus improving the applicability of this contextual approach.

3

Central China from the Late Neolithic to the Early Bronze Age: an Archaeological Background

This chapter begins with a brief introduction of the cultural succession in the Central Plains region from the late Neolithic to the early Bronze Age, with focus on dating and geographical distribution of sites. Production, subsistence, settlement pattern and social organisation of each culture/period are organised in a comprehensive synthesis. Having presented this panorama of the general understanding of this period, I follow up with a critical review of the scholarship on the issue of early states in China. By going through these facts and arguments, I aim to provide an overview of this period in material culture as well as scholarly interpretation and hence clarify the constraints imposed by the nature of archaeological data and inadequacy of the current theoretical framework.

3.1 Archaeological information

The Central Plains region, situated in the middle Yellow River basin, covers mainly modern Henan Province, the south part of Shanxi and Hebei Province and the west part of Shandong Province. The region forms the central focus of this book. Its rich archaeological background opens up the arena for the contextual discussion of the animal categories in the Central Plains.

From the late Neolithic through the early Shang dynasty, four archaeological cultures/periods constitute the chronological sequence in the Central Plain: Henan Longshan culture or Wangwan III phase (*ca.* 2600~1900 BC), Xinzhai culture (*ca.* 1850~1750 BC), Erlitou culture (*ca.* 1900~1500 BC) and Erligang culture (*ca.* 1600~1400 BC). **Figure 1.2** in chapter one reveals the chronology for the research period and relevant dates for our three study sites. The three sites span the Longshan period, Xinzhai and Erlitou period, and end around the Erligang period. These four periods are clearly indicated by four different types of material culture characterised by their different ceramic types. A possible change in the patterning of archaeological evidence and the distribution of sites was found between the Erlitou and Erligang period, indicated in the shifting ceramic style, which some scholars have interpreted as indicative of a political reshuffle between late Neolithic societies and the state-level polity (Liu & Chen 2006: 162-8), although other archaeologists (*e.g.* Shelach & Jaffe 2014) remain skeptical about such a supposed cultural gap.

3.1.1 Late Longshan period

The late Neolithic Longshan culture in the Central Plains is usually termed the Wangwan III phase or Henan Longshan culture in order to differentiate it from other local variations, such as Shandong Longshan culture in Shandong Province, or Shanxi Longshan culture (*i.e.* Taosi type) in Shanxi Province. The focused period in this book is roughly within the temporal frame of the late Longshan, dating to *ca.* 2600~1900 BC (Institute of Archaeology, CASS 2010: 535). Over thirty late Longshan sites have been excavated in the Central Plains (**Figure 3.1**), among which Wangwan (Yan & Li 1961), Wangchenggang (HPIACR & MCH 1992), Meishan (Yuan 1991) and Haojiatai (HPIACR 2012) are representative sites (Institute of Archaeology CASS 2010: 532-3).

Typologically, the Henan Longshan culture is characterised by *guanxingding* (jar-shaped tripod), *aizuding* (short-legged tripod), *jia* (tripod drinking cup) and a few of other types of dining and drinking vessels formed of grey clay (Institute of Archaeology CASS 2010: 533). The dedication manifested in these ceramic vessels is indicative of a standardised pottery production (Underhill 1991; Liu 2004; Institute of Archaeology CASS 2010: 540). It seemed that metallurgy was practiced on a small scale, as pieces of copper fragments were found in a limited number of sites (Institute of Archaeology CASS 2010: 540).

A number of walled sites emerged, around which smaller settlements were clustered, forming a two-tiered hierarchical settlement pattern in most regions (Institute of Archaeology CASS 2010: 537). In the study region of the modern-day Henan Province, the distribution of site-clusters even reveals a more complex three-tiered settlement hierarchy (Liu 2004: 182-3; Zhao 2001). Several fortified sites have been excavated in this region, including Puchengdian (Wei 2008), Pingliangtai (Cao & Ma 1983), Wangchenggang (HPIACR & MCH 1992), Haojiatai (HPIACR 2012), Guchengzhai (Cai *et al.* 2002), Xinzhai (ACSACPU & ZMIACR 2008), Mengzhuang (Yuan 2000) and so on. Each was probably a major centre in their settlement clusters, surrounded by secondary centres and lower ranking small villages (Liu 2004: 183-5). Within those walled sites, large buildings were found, though their function remains controversial (Shelach & Jaffe 2014), with interpretations ranging from palace (HPIACR 2002; Liu 2004: 104–5) to public hall (Li 2010). In addition, some smaller buildings might have been employed as residential structures (Institute of Archaeology CASS 2010: 539).

Underneath or near buildings, 'foundation depositions' were frequently recovered containing single or multiple human or animal skeleton(s) (*ibid.*: 541). Similar

Figure 3.1 Map showing the distribution of late Longshan sites in the Central Plains (1. Wangwan, 2. Meishan, 3. Haojiatai, 4. Puchengdian, 5. Pingliangtai, 6. Guchengzhai, 7. Mengzhuang sites are mentioned in the study)

depositions were also found under or close to walls in some fortified sites, such as at Wangchenggang (HPIACR & MCH 1992). They were commonly interpreted as remains of ritual performances associated with construction (Institute of Archaeology CASS 2010: 541). Other activities interpreted in some literatures as ritual practices include certain forms of oracle divination (*ibid.*). Oracle bones, sometimes with scorch marks, were collected from sites such as Miaodian, Xiaopangou, Haojiatai and Wadian. Most all oracular bones were made from scapula of pig, cattle, sheep and deer (*ibid.*).

Only a handful of cemetery sites have been discovered, rendering limited insight into mortuary practice. As reported, up to 300 tombs have been excavated in this region to date, mostly child urn burials. While adult skeletons were sometimes found buried within grave pits, hardly any grave goods were recovered (*ibid.*: 542). The general impression received is that funeral practices in the region do not demonstrate any obvious social polarisation. Beyond the local scope, however, evidence from other regions, such as cemetery sites at Taosi, Chenzi, Yinjiacheng and Zhufeng, does indicate the existence of a kinship-based stratified social structure during the Longshan (Liu 1996).

People might have already adopted an agricultural lifeway mainly reliant on millet crops and supplemented by rice cultivation (Institute of Archaeology CASS 2010: 540). Animal husbandry could have formed another vital aspect of subsistence. The major types of livestock include dog, pig, cattle and sheep. Details about animal husbandry will be discussed in a later section.

3.1.2 Xinzhai period

The Xinzhai period, named after the discovery of the Xinzhai site in the Xinmi city, Henan Province, is generally considered the transitional period between the late Longshan and early Erlitou, and dates from the latter half of the nineteenth century BC to the first half of the eighteenth century BC (Zhang *et al.* 2007). As reflected in material culture, the types of pottery frequently excavated and the combination of ceramic vessels are distinctive. Using these material cultural criteria, it has been found that the distribution of Xinzhai culture sites was restricted in a very small region running east to Songshan including modern-day Zhenzhou, Gongyi, Xinmi, Xinyang and Xinzhen, a spread that overlapped with the northeastern section of the Henan Longshan culture (ASCAS & ZMIACR 2008: 531). Only a small number of sites yielded remains from the the Xinzhai period: Erqilu (HPIACR 1983), Niuzhai (HPIACR 1958), Mazhuang (Li & Liao 1982), Huangzhai (HPIACR 1993a), Shuhe (HPIACR 1996) and Huadizui (Gu & Zhang 2003). This type of material culture was soon replaced by Erlitou culture. Considering the limited geographical distribution and short chronology, it appears that Central Plains, the transition from late Longshan to Erlitou took different local trajectories. Late Longshan culture transformed directly into the Erlitou culture in certain locales, as exemplified by Wadian and Wanghenggang, whereas in the eastern Songshan area late Longshan culture was followed by Xinzhai period and then finally replaced by Erlitou culture (ACSAC & ZMIACR 2008: 540-1). A more sophisticated understanding of the transitional Xinzhai culture requires new evidence.

3.1.3 Erlitou period

Erlitou culture remains were first discovered in 1957, since which time excavations have continued for decades. The Erlitou culture spans the period from 1880 BC to 1520 BC (Xia-Shang-Zhou Chronology Project Team 2001). The majority of the sites are distributed in the central and west Henan Province, the Luoyang Region and southwest Shanxi, the first of these areas falling within the study region in this book (**Figure 3.2**). Within the study region, a considerable number of sites with Erlitou remains have been reported: Erlitou (Xu 2004), Huizui (HPIACR 1990; Chen *et al.* 2010), Wangchenggang (HPIACR & MCH 1992), Meishan (Yuan 1991), Xinzhai (ACSAC & ZMIACR 2008), Donggangou (Institute of Archaeology CASS 1958), and so on. Among them, the Erlitou at Yanshi, Luoyang, is considered to have been the economic and political centre of Erlitou Culture. Erlitou ceramics can be found in many other regions, reaching east Shanxi, north Hubei and east Henan (Institute of Archaeology CASS 2003: 82-3).

Metal production gradually became specialised during the Erlitou period. A remarkable number of bronze vessels have been excavated, focused around the Erlitou site (Institute of Archaeology CASS 2003: 109). Metal foundries were also discovered at the Erlitou and Dongxiafeng sites (Institute of Archaeology CASS *et al.* 1988). This specialised metal production technology was under the rigid control of political elites and the distribution of metal products as prestige goods and ritual objects was likely restricted among elites of high social status (Liu & Chen 2006). Jade production was also highly skilled and well-organised.The tradition can be traced back to the middle Neolithic, particularly in coastal regions (*i.e.* in Hongshan

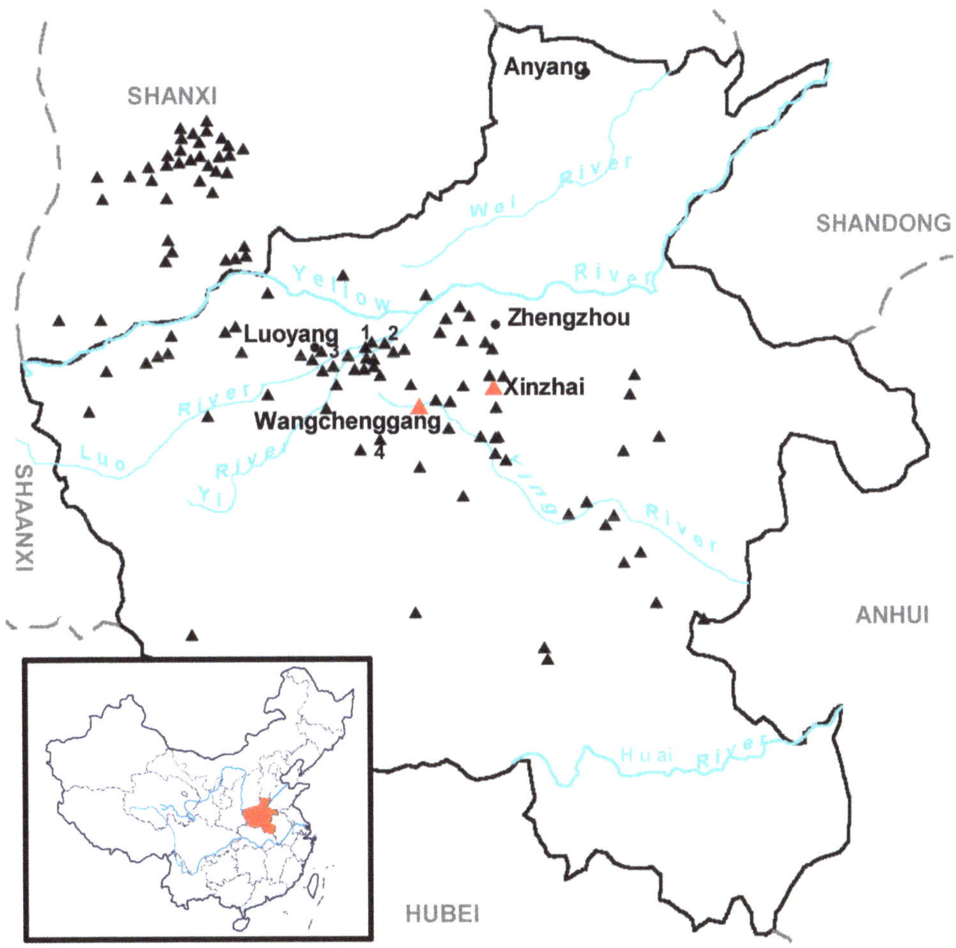

Figure 3.2 Map showing the distribution of Erlitou sites in the Central Plains (1. Erlitou, 2. Huizui, 3. Donggangou, 4. Meishan sites are mentioned in the study)

culture) (Dematte 2006), but both the scale and skill improved markedly during the Erlitou period (Institute of Archaeology CASS 2003: 116-7). In many archaeological narratives these jade items are described as prestige goods and ritual objects (*e.g.* Institute of Archaeology CASS 2003). Potters appeared to have formed a group of specialised craftsmanship in the central Erlitou site (Institute of Archaeology CASS 2003: 120).

Compared to the previous Longshan period, a more complicated settlement pattern comprising of a four-tiered hierarchical ranking emerged in the Erlitou period (Liu & Chen 2003). The emergence of this settlement hierarchy was accompanied by the occurrence of the super-size Erlitou site as the primary centre among contemporaneous sites (Liu & Chen 2003; Liu 2004: 183). The Erlitou centre covered an area of 300 ha during its heyday. The palatial complex, located around the centre of the site, was enclosed by a rammed-earth wall cutting it off from the ordinary residential and burial area west and north, respectively. The ceremonial area, characterised by ritual structures and burials, was situated north of the palatial complex, and bronze workshops lay to the south (Xu 2004; Liu 2009).

Our understanding of Erlitou mortuary pattern is based upon the 400 tombs excavated so far. 300 of these burials lay within Erlitou. The quality and quantity of grave goods along with the size of grave pits suggest social stratification (Liu & Chen 2006). Nonetheless, no long-term cemetery site was discovered. Instead, burials and houses occupied the same area alternately, so we lack for any clear thread that would demonstrate kinship (Institute of Archaeology CASS 2003: 101). Liu and Xu (2007) propose that the absence of long-term cemetery sites was likely the result of rapid population migration.

Much weight was put upon ceremonial rites in the Erlitou culture. Particulary at the Erlitou site, special types of structures relating to ritual ceremonies were aligned north of the palatial complex (Institute of Archaeology CASS 2003: 129). These types of buildings were associated with *tan* 坛 (raised platform) and *shan* 墠 (shallow pit), mentioned by several textual passages composed in later dynastical periods, which albeit depict early Bronze Age rituals according to their own understandings (*ibid.*: 130).

In addition to the discovery of the mega-site at Erlitou, a considerable number of small villages are found scattered over the Central Plains. Liu (2004: 236) speculates a core-periphery network for the Erlitou, where the centre played its role in production while the surrounding villages of rich raw materials did their job in provision. What is more, this inter- and intra- regional network was established and maintained by military conquest and control from the Erlitou centre (*ibid.*: 232-4).

3.1.4 Erligang period

The Erligang period represents the early phase of the Shang dynasty and dates to 1600~1400 BC. The period is recognised by ceramic typology (Xia Shang Zhou Chronology Project Team 2000; Institute of Archaeology CASS 2003: 188). Erligang culture sites were distributed over an extremely wide area that includes modern-day Shanxi, Hunan and Shandong Provinces. The majority of sites were distributed within the Yiluo basin (**Figure 3.3**). Erligang culture remains have been found in many archaeological sites including Yuanqu Shang city (MCH 1996), the Yanshi Shang city (Duan *et al.* 1984), Shaochai (HPIACR 1993b), Wangchenggang (HPIACR & MCH 1992), Shangjie (Wang *et al.* 1966) and Qilipu (ATYRR 1960).

Cultivated crops and domestic animals were still the main food sources during the Erlitou period (Institute of Archaeology CASS 2003: 107). Millet and rice were two major staples, macro-remains of which were found in some sites, while cattle, pig, sheep and dog comprised the majority of faunal remains. Apart from animal husbandry, exploitation of wild animals might have been frequently practiced, augmented by the presence of substantial deer bones (*ibid.*: 108-9). Animals also provided raw materials for a bone industry in addition to food resources. Bone tools were widely found among Erlitou period sites. At the Erlitou site the substantial accumulation of bone materials for tool production indicates the potential site of a bone workshop (Institute of Archaeology CASS 2003: 121). In this workshop, one of the bone tools was carved with '*taotie*' (a mythical beast) pattern on it, which serves as a good example of the craft proficiency (*ibid.*).

Bronze production became more sophisticated during the early Shang. Mold-casting technology became capable of manufacturing complicated multi-component ritual vessels (Institute of Archaeology CASS 2003: 373-6). The number of metal foundries increased, most of them still located adjacent to walled towns such as the Zhengzhou Shang city (HPIACR 1989). Ritual and funeral vessels occupy a large proportion in the bronze assemblages excavated and their distribution was clustered around Zhengzhou, Yanshi and other fortified mega-sites (Institute of Archaeology CASS 2003: 379). Pottery production remained a specialised activity and vessel types became more standardised (Institute of Archaeology CASS 2003: 404-8). Jade objects were found in several walled sites, but decreased in number and diversity. It seems as if bronze objects replaced jades as prestige goods of symbolic significance (Institute of Archaeology CASS 2003: 408).

It has been generally accepted that the Erligang settlement hierarchy covered a wider range of regions and involved at least a four-tier ranking. The Zhengzhou and Yanshi sites ranked in the first class as primary centres of the Shang polity, while beyond the central region, the Dongxiafeng and Panlongcheng seemed regional centres ranked in the second class (Institute of Archaeology CASS 2003: 236). The third and forth classes include smaller villages scattered round those large settlement (*ibid.*). There are shared traits to these fortified sites. Firstly, they were generally rectangular, modified to adapt to the local

Figure 3.3 Map showing the distribution of Erligang sites in the Central Plains (1. Dongxiafeng, 2. Yanshi, 3. Shaochai, 4. Shangjie, 5. Qilipu, 6. Yuanqu sites are mentioned in the study)

geography. Secondly, settlements were fortified by rammed-earth walls. Thirdly, the palatial complex was located either in the north part of the site, separated from commoners' habitation and production areas, or in the centre of the settlement along with the craft production workshops. Fourthly, storage structures were recovered within sites (Institute of Archaeology CASS 2003: 237).

So far, no large-scale cemetery ground has been found near these fortified settlements, other than a few of individual tombs scattered within residential and workshop areas (Institute of Archaeology CASS 2003: 237-8). The variety of tomb size, along with the varying quantity and quality of funeral items has been perceived as a reflection of a stratified social order (Institute of Archaeology CASS 2003: 242).

Oracle bone inscriptions enter the equation during the late Shang Dynasty. The existence of the writing system in the early Shang remains in doubt due to the lack of reliable evidence. Dozens of pottery marks have been found dated to the Erligang period and earlier (Pei 1993). There were also occasional discoveries of bone marks resembling the oracle bone inscriptions of the late Shang, but the date has not yet been confirmed (Institute of Archaeology CASS 2003: 423-4).

The agricultural lifeway remained prominent during the Erligang. Millet was the major crop, followed by wheat, likely a newly introduced cultivar (Lee *et al.* 2007). A large collection of rice remains was recovered from Yanshi (Institute of Archaeology CASS 2003: 372). As rice remains were scarce at the smaller village sites, it is possible that rice might have been a valuable commodity during this time. On the other hand, he Yiluo basin provided ideal environment for rice production, so the possibility that rice was adopted as an ecological adaptation cannot be easily dismissed (Lee *et al.* 2007). Animal husbandry, without doubt, made a salient contribution. Inferred from oracle bone inscriptions in late Shang, domestic animals might have also played a vital role as sacrifices to ancestors (Institute of Archaeology CASS 2003: 372; Yuan & Flad 2005). Royal/elite hunting was likely also practiced (Institute of Archaeology CASS 2003: 372).

3.1.5 Research period: an overview

On balance, the late Longshan Central Plains population was skilled skillful in types of craft production like pottery production, and had mastered farming. Social networks began to involve several regional sites, with one or two sites sometimes fortified and always serving as the regional hub. Increasing violence and conflicts accompanied the increase in population. The Bronze Age begins with the Erlitou culture, when social hierarchy is strengthened, reflected in the remarkable architectural differentiation and elaboration of differing burial types between elites and commoners. Bronze production became sophisticated and specialised, focusing on casting ritual vessels for those with high status. The Erligang period marks the conception of the Shang dynasty. The scale of the central city expanded dramatically and social stratification and segregation were intensified, as evidenced in the layout of the mega-site. The settlement network also extended to cover a larger region while engaging more sites of different ranks within a tighter centralised control. This highly centralised power permeated multiple aspects of daily life and included control over prestige goods (metal), violence (hunting and war) and the unknown (divination and sacrifice). The centralisation of the power was also manifested in the massive area and intense population under control. Social hierarchy became more complicated where the differentiation among elites began to appear. Following this trajectory, the late Shang finally developed into an extremely hierarchical, centralised society.

Our three case studies fit into the cultural sequence successively (**Figure 1.2**). The Wadian site covers the late Longshan. The date of Wangchenggang overlaps the latter half of the sequence at Wadian. Though experiencing many episodes of reconstruction, the Wangchenggang site prolonged from the late Longshan through Erligang periods. As demonstrated above, a hierarchical settlement system had been established since the Longshan. A series of features displayed by the site indicates its central place within the hierarchy. This central importance, however, was substantially reduced during the Erligang period when the political seat shifted northeast to Anyang. Among the three sites, Xinzhai has the shortest period of occupation, beginning at around the end of the Longshan and terminating around the Erlitou. In sharp contrast to the short chronological sequence, Xinzhai seems to play a central role in the regional network, as evidenced by a long list of distinctive characteristics including multiple fortified walls, the elaboration of bronze vessels, and so forth. More detailed information of each site is enclosed in the case studies presented in chapters four, five and six respectively.

3.2 Archaeological interpretation: terminology and approach

The period spanning the late Longshan through early Shang has long been a research focus central for exploring the origins of the Chinese civilisation or, in alternative terminology, the early Chinese state. A plethora of studies have contributed to examining this issue, ironically much bitter controversy rather than helping to resolve the question. Various methods of interpreting archaeological evidence and correspondingly varying understandings of archaeological phenomena are grounded in distinct research theories and approaches. It is helpful to step back and critically review the scholarship and the understanding of the past forged in each case.

Scholars often employ a trio of terms to describe the sociopolitical condition of this period; they are 'civilisation', 'dynasty', and 'state'. These terms 'civilisation' and 'state' are commonly used interchangeably without applying a rigid distinction (Liu 2009). The legendary Xia dynasty appearing in the historical narrative is archaeologically correlated to the state-level Erlitou culture from which the incipient Chinese civilisation grew. In this way the three terms are intertwined and bring the research into the limbo of ambiguity. Terminology, though subject to semantic adjustments, in fact will be consistent with the approach where it has been employed. The three terms here demonstrate three distinct methodological frameworks developed from various social backgrounds to approach the same issue differently.

3.2.1 Dynasty and historiographical tradition

The use of the term 'dynasty' reveals the obsessions with the historical paradigm within Chinese archaeology (Falkenhausen 1993). 'Dynasty' is one of the most frequent chronological units in historical discourse. In multiple written documents such as *Shiji 史記 (The Scribe's Records)*, *Xia 夏* is the first Chinese dynasty, founded by *Dayu 大禹* (the Great Yu). After the establishment of archaeology as a modern discipline in China during the early twentieth century, Chinese archaeologists felt the urgency to confirm the existence of the Xia Dynasty archaeologically, motivated by the search for the indigenous origins of Chinese culture (Liu & Chen 2012: 6; Shelach & Jaffe 2014). Under such academic ambitions and political motivations was the archaeological research of the Xia launched. A handful of archaeological expenditures (*e.g.* Xu 1959) and excavations (*e.g.* Institute of Archaeology CASS 1999) were conducted that adopted this historical-geographical and, accordingly, scholars would claim that the Erlitou site was the capital of the Xia dynasty. More recently, the Xia-Shang-Zhou Chronology Project, employing archaeometric methods to build up a systematic and reliable chronology of the Three Dynasties (Li 2002), though producing newer and more accurate dates, has been criticised as remaining shackled to the historiographical framework stressing dynastic succession and over-reliant on textual records (Lee 2002; Shaughnessy 2009; Shelach & Jaffe 2014).

Textual documents in China have recorded a long time span in considerable detail. The language employed in these texts remains identifiable nowadays. These features make ancient texts a source of reference to search for a progressively earlier past. Accordingly, an approach integrating ancient texts, geography and archaeology has been formulated. This can be seen as a Chinese variant of direct historical approach (as defined by Wedel 1938). One example of the successful application of this historical-geographical approach was Xu Xusheng's discovery of the Erlitou site in 1959, achieved by tracking down ancient texts to pinpoint the exact location of *Xiaxu 夏墟* (the ruins of Xia) (Xu 1959) . Despite the heated debate disputing whether the Erlitou site was the capital or whether the Xia dynasty existed in essence, it remains in no doubt that Erlitou was indeed one of the most significant archaeological site in the region and that its first discovery was aided by the reading into ancient texts. From a methodological standpoint, the example demonstrates the potential for archaeologists to critically extract information from ancient written records and integrate them into the study of the past.

3.2.2 Civilisation and nationalism

Most research which adopts the term 'civilisation' busies itself with compiling lists of traits believed to be signals of the inception of Chinese civilisation. Xia Nai was the first Chinese archaeologist who adopted the term 'civilisation' in an academic sense in order to denote the nature of the Erlitou culture (Xia 1985: 81). His understanding of 'civilisation' was considerably influenced by Childe's view on early civilisation and urban revolution (Childe 1950, 1957). In addition to its academic meaning, the use of 'civilisation' boasts the nationalism under which archaeology was employed as a powerful tool to justify the originality and long duration of Chinese civilisation, and leave it inferior to no other civilisations. Enraptured by this nationalistic tide, some scholars even traced the beginning of 'civilisation' back to the late Neolithic in the fourth millennium BC, a thousand years or so earlier than the current understanding (Falkenhaussen 1993).

Archaeology of this time worked efficiently to stress the continuity of Chinese civilisation and thus assured the legitimacy of the modern political entity. Such efforts have not been entirely abandoned. Recently, the national-level 'Origins of Chinese Civilisation' project has been launched, and one of its aims is to archaeologically authorise the existence of the Xia. More than often, the Xia is still seen as a legendary period lacking concrete evidence of existence. Archaeological studies keen on seeking for 'civilisation', consequently, carry this tag of political significance.

3.2.3 State and neo-evolutionary models

The term 'state' is usually used in studies that employ neo-evolutionary models, exemplified by a series of settlement studies conducted by Liu and colleagues (*e.g.* Liu 2009; Liu & Chen 2012). The foundation for this approach lies in models of cultural evolution which was initially established by Fried (1960) and Service (1975), and advocated for more or less universal laws of social development. As these laws and principles are considered universal, such models can be applied cross-culturally to archaeological sites in China. In this approach, Erlitou culture is recognised as a 'state' according to a set of criteria: 1) the existence of extremely large site as the state centre; 2) a four-tier hierarchical settlement pattern; 3) military expansion and control over a broad swathe of territory; 4) a site layout reflecting social segregation; 5) specialised craft production of prestige goods under the centralised control; 6) sociopolitical stratification manifest in mortuary practices (Institute of Archaeology CASS 1999; Liu & Chen 2003; Liu 2004; Liu 2006; Liu 2009).

Settlement analyses were initially applied to late Neolithic China as a scientifically guided research substitute of the traditional cultural-historical approach focused on searching for legendary capitals of the Three dynasties vaguely mentioned in ancient texts. Such applications can be seen in works by Chang (1983), Liu & Chen (2001), and others. These studies have contributed to our understanding of the late Neolithic as an era of multiple cultural interactions involving a variety of aspects instead of a lineal chronological changes in political powers. However, the approach is first and foremost criticised as imposing an over-generalised anthropological observation onto specific archaeological records by sneakily substituting the idea of settlement with the concept of certain political entity (Wolf 1982; Trigger 1998; Yoffee 2005). Secondly, by presuming a certain universal model, it downplays social developments by reducing the progress into a more or less uniform model subject to a set of uncritical traits (*ibid.*).

Despite all the opposing voices, this approach succeeds in ridding one of the traditional historical paradigm and thus settling the study of Erlitou at a worldwide archaeological avenue where cross-cultural communications can become possible.

3.2.4 Relational studies: an alternative

Whatever the title applied – dynasty, civilisation, state – all such approaches in fact strive to search for the landmark signaling a vital change in the course of Chinese history. Nevertheless these theoretical frameworks themselves are comprised of anachronistic conceptions detached from the original context in the past and the role of agency is invariably muted in pursuit of the larger picture. To avoid controversies in working definitions and dating, some archaeologists decide on a relational study focusing on either nested relationship among people and people, people and things within the society (Campbell 2009) or relative changes in the society along the same trajectory through time (Shelach & Jaffe 2014). Such studies have rendered remarkable insights and pointed out a new direction to address the relevant issues.

While the controversy remains unresolved with regard to conclusions as well as approaches , it is generally accepted that the Erlitou period was the turning point when a broad view of early China is taken, as before and after Erlitou, complex societies, likely with social stratification, craft specialisation and labour division at various levels from the terminal Neolithic shifted to an extremely hierarchical society under the control of a strong institutional power in the late Shang. Nonetheless, this might have been a gradual change, meaning we could not demonstrate clear signs of an exact beginning and end. Changes might also have involved a dynamic interplay of a variety of factors, one which can hardly be covered by a single theoretical framework. Different studies take different perspectives.

In short, here I prefer to conservatively adopt Shelach & Jaffe's (2014) conclusions from their long-term trajectory approach: two phases of rapid transformation were identified, the late Longshan period around 2500 BC, and the Erligang period around 1600 BC. The research period for this project lies exactly in-between. As the research focus — human and animal relationship — itself is dynamic and interactive, instead of focusing on the identification of 'state' 'civilisation' or other, I advocate Campbell' s (2009) networks and boundaries approach, that is, viewing society as a network composing of people, objects, events and so forth, with interactions among them. I attempt to contextually explore the issue of animal classification against its original economic, social and cultural background.

3.3 Animals at the archaeological sites

Faunal assemblages excavated from sites in the Central Plains demonstrate a diversity of animal resources exploited in the past. They reflect not only natural abundance and diversity of animal populations, but were also filtered by human selection and alteration. The following section is a general introduction of animals frequently found in Central Plains sites dating from the late Neolithic through initial Bronze Age, focusing on deer, dog, pig, cattle, sheep/goat and horse. Their origins, interactions with people and roles in the economy and society will be summarised.

*3.3.1 Deer (*Cervidae*)*

21 taxa of wild deer are distributed around modern-day China (Sheng *et al.* 1992: 8). Central China experienced a warm and mild climate from 5000 BC. As a consequence, many species of *cervidae* for which the modern distribution is restricted to southern China once lived in northern and central regions, for example sika deer (*Cervus nippon*) and Père David's deer (*Elaphurus davidianus*)(Sheng *et al.* 1992: 267).

Deep exploitation was consistent from the Palaeolithic to the Bronze Age. Venison might have provided good sources of protein. In addition, deer long bones and antlers were also useful raw materials for tool and ornament production, commonly seen in archaeological assemblages.

Deer could have been the ideal protein resource prior to animal domestication, being generally abundant and relatively easy to capture. The substantial volume of deer remains excavated from sites before or at the initial stage of the domestication supports this postulation. Although there was a tendency for deer remains to decrease in response to the increasing utilisation of domestic animals, this change does not seem to have been uniform cross-regionally. In the Jiangzhai and Kangjia sites there were both resurgences of deer exploitation during the late Neolithic (Qi 1988; Liu *et al.* 2001). During the Shang, deer gained increasing importance as the prey of royal hunts. Given the large numbers recorded in oracle bone inscriptions, an enormous amount of deer must have been hunted (Zhu 2005).

It remains up for debate whether manipulation of wildlife to a certain extent had been conducted in the past. Jiangzhai and Xinzhai are two sites where deer management has been proposed, based on age profile and isotope data (Zhang & Zhao 2015). This issue will be addressed in chapter six.

*3.3.2 Dog (*Canis lupus familiaris*)*

It is widely agreed that the dog was the first domesticated animal (Clutton-Brock 1995; Serpell 1995; Vila *et al.* 1997; Larson *et al.* 2012). The detailed process of this domestication, however, remains ambiguous.

The same can be said of dogs in China. We undersood little about the domestication process in terms of ancestry, exact time and location of the domestication and number of domestication events. Genetic evidence, though, depicts the genealogy in detail, results of various studies employing different samples and methods are extremely contentious (Vila 1997; Savolainen *et al.* 2002; Lindblad-Toh *et al.* 2005; Pang *et al.* 2009; Vonholdt *et al.* 2010; Skoglund *et al.* 2015). The earliest morphologically domestic dog was found at Nanzhuangtou in Hebei Province *ca.* 8000 BC (Yuan & Li 2010). In other early Neolithic sites such as Jiahu (HPIACR 1999), Cishan (Sun *et al.* 1981) and Dadiwan (Qi *et al.* 2006), dogs were largely found deposited in burials or pits near houses, with complete skeletons. Dogs seem to have been treated differently from other food animals like the pig. The distinct treatment of dogs is commonly interpreted as reflecting their proximity to human beings as hunting assistants (Liu & Chen 2012: 98).

Dogs were also consumed. Dog remains have been found deposited in refusal pits at many sites, spanning from the Neolithic to the Bronze Age.

In the lower reaches of the Yellow River, Shandong Province in particular, the dog gradually ranked as a sacrificial animal with ritual significance (Gao & Shao 1986). This institution seems to have been inherited by late Shang society in the Central Plains, where dogs became the routine animal sacrifices for mortuary practices (Li 2011).

*3.3.3 Pig (*Sus scrofa domesticus*)*

The wild boar (*Sus scrofa*) is indigenous to China as recorded in fossil remains from the Pleistocene. Wild boars have been exploited since the Palaeolithic, as displayed in faunal assemblages found in many sites across various regions (Yuan & Flad 2002). Accessibility to the wild boar is proposed to have contributed to the domestication of pig. North China was most likely an early centre of pig domestication. The earliest evidence of domesticated pig in central China is tracked back to 6600 BC. at the Jiahu site, in Henan Province, and is based on dental geometric morphometrics (Cucchi *et al.* 2011), age profile and archaeological context (Luo & Zhang 2008). The earliest domestic pig might have been derived from local wild swine, as a north/south variance in size among domestic herds corresponds to geographical size differentiation in wild boar during the Pleistocene (Luo 2012: 243-4).

The Neolithic witnesses a trend for increasing intensity of pig domestication across the Central Plains (Yuan & Flad 2002; Dong & Yuan 2020). During the middle Neolithic, at around the fifth millennium BC, the subsistence economy had become heavily reliant upon domestic pigs (Liu & Chen 2012: 104). Pigs also seemed to bear symbolic weight as supplies for feasting and rituals. At Baligang in Henan, for example, pig mandibles were one of the most popular funeral items, and found in substantial number (Zhang *et al.* 2000). There remain exceptions, particularly in the lower Wei River region, such as at Jiangzhai and Kangjia, where pig bones decreased considerably accompanied by an increase in wild fauna (Qi 1988; Liu *et al.* 2001; Liu & Chen 2012: 104). While pigs were still on the menu for Bronze Age people, their contribution seemed to gradually reduce following introduction of grazing animals from West Asia (Luo 2012: 379). Cattle, sheep and horses also mostly substituted the ritual role during the Shang (Yuan & Flad 2005).

Our understanding of pig husbandry is largely constrained by the resolution of archaeological data. From the broadest point of view, initial domestication was most likely completed about 9000 years ago, since which time the pig played a salient role in both subsistence and symbolic domains. With the introduction and rearing of new grazing species such as sheep, cattle and horse, reliance on pigs was slightly relieved, though retaining its contribution to human subsistence. When the picture is magnified, however, these husbandry strategies have been little explored in actual depth. Where isotopic data are available, pigs' consumption of a C3 diet in the early Neolithic shifted to C4-based intake in the middle and late Neolithic and Bronze Age, suggesting increasing rigid supervision of domestic herds along with a process of privatisation (Liu & Jones 2014). No further chronological change is readily observed from the late Neolithic to the Shang, reflecting the stability of pig husbandry strategy through the time in the Central Plains (Dai *et al.* 2015). Recently isotopic and dental geometric morphometric results of pig remains at Xiawanggang and Xinzhai demonstrate a pair of husbandry strategies before and after the Neolithic—Bronze Age transition: one associated with the household management, resisting an expanding centralised authority, and the other the extensive animal raising strategy under the supervision of elites (Cucchi *et al.* 2016). The speculated two models, however, await the testimony of more extensive data from other sites.

*3.3.4 Cattle (*Bos taurus*)*

Cattle were most likely introduced into the middle and lower Yellow River region around the third millennium BC (Yuan *et al.* 2008). Several depositions of complete or articulated cattle skeletons were found in late Longshan sites, such as Shantaisi (Yuan 2010) and Pingliangtai (Cao & Ma 1989) in Henan. The beginning of the Shang dynasty sees an increasing amount of cattle remains excavated from archaeological sites. Measurements on these cattle bones provided morphometric confirmation of their domestic status (Lü 2007: 11-2). Genetic studies show that both taurine (*Bos taurus*) and zebu (*Bos indicus*) cattle contributed to the gene pool of modern Chinese cattle, with taurine mtDNA recognised in the northern breed and zebu mtDNA in the southern breed. Central Plains cattle might have derived from a hybrid form (Lai *et al.* 2006; Cai, X. *et al.* 2007). It seems plausible that two different species of cattle arrived in China via northern and southern different routes and may have admixed in the Central Plains (Liu & Chen 2012: 106).

Cattle only account for a minor proportion of Longshan assemblages (Yuan 2010). Around a century later in the Shang dynasty, cattle became multi-usable and highly valuable, providing raw material sources for bone artifact production in addition to providing a considerable source of meat. The discoveries of large-scale bone workshops at the Great Shang Settlement at Anyang support this argument (Ma 2010; Campbell *et al.* 2011).

Aside from their pragmatic utility, cattle also played a vital role in ritual ceremonies. Cattle were ranked as precious sacrificial chattel, especially during the Shang dynasty, as manifested in both archaeological depositions (Yuan & Flad 2005) and ancient textual passages such as the *Liji* 禮記 (*Book of Rite*). The ritual weight allocated cattle, however, seemed to have deep historical roots, beginning at Erlitou at the the latest when a systematic and standard husbandry strategy of cattle involving year-round millet-byproduct foddering might have been practiced to ensure the proper management of cattle, as illustrated by the isotopic shift from a relatively broad dietary spectrum to a restricted intake pattern in cattle samples from several late Neolithic sites (Dai *et al.* 2015; Dai *et al.* 2016). This specialised management of cattle, probably under some elite power, is postulated to correspond to the process of urbanisation and state formation (Dai *et al.* 2016).

*3.3.5 Sheep and goat (*Ovis aries *and* Capra hircus*)*

I have included sheep and goats together due to the difficulty of morphological distinction in archaeological samples first and foremost. Many archaeological reports involved in this study do not distinguish sheep from goat. Secondly, when they are distinguishable, sheep remains substantially outnumber those of goats. Especially when the study region is taken into account, the Central Plains environment was better adapted to sheep rather than mountain-adaptive goats.

Wild species of sheep and goat did occur in several early Neolithic sites but the earliest remains of domestic sheep (*Ovis aries*) have been found concentrated in west China and Inner Mongolia, such as the Majiayao site in Gansu and at Hongshangou in Inner Mongolia in Neolithic sites from around the fourth millennium BC (Zhou 1984). New radiocarbon dates and aDNA results of ovicaprid remains from the Tao River Valley in Gansu, however, suggest a much later introduction (Brunson *et al.* 2020). No concrete conclusion can be achieved about the route of transmission. It is not until the Longshan period that sheep along with goats (*Capra hircus*) began to appear in the Central Plains. Although China contained local wild caprines, domestic sheep might have been introduced rather than indigenously domesticated. Genetic evidence suggests the Near Eastern origin of Chinese domestic sheep (Guo *et al.* 2005) , while an extinct lineage with unknown geographic origin might also have contributed to the modern domestic population (Cai *et al.* 2007; Cai *et al.* 2011).

Sheep and goats might have been introduced into China across the Eurasian steppe, paralleling the spread of wheat and barley from the west across an as-yet unclear route (Flad *et al.* 2007; Zhao 2009). The eastward movement of domestic sheep and goats along with crops forms a picture of food globalisation in the prehistory (see Jones *et al.* 2011). In addition to these cultural and economic contacts between traditional farmers and steppe pastoralists, the middle Holocene climatic optimum could have also aided the introduction and spread of sheep and goats (Liu & Chen 2012: 106). Agropastoral subsistence was first established in the Ordos Region at the end of the third millennium BC and less than a few centuries years later it became widely adopted in north China by the early half of the second millennium BC latest (*ibid.*).

Sheep and goats, with little doubt, were raised for the single purpose of providing meat. An increasing amount of evidence for kill-off pattern points to the exploitation in secondary products, probably wool in the Central Plains, exemplified by the Xinzhai and Erlitou sites (Li *et al.* 2014; Dai *et al.* 2015). Sheep were also sacrificed during the late Shang. Their ritual importance even surpassed traditional pig sacrifice, ranking just below the sacrifice of prestigious cattle (Yuan & Flad 2005).

Only a handful of studies have attempted to demonstrate sheep husbandry strategies in detail. Isotopic data offer a line of evidence. Late Longshan husbandry possibly involved seasonal supply of millet-contained fodder either through direct feeding or letting herds graze the cropland after harvest (Dai *et al.* 2016). The surmise, however, is based upon on the analysis of one single site and thus awaits testament from other sites. Few studies focus on Erlitou and Erligang sheep husbandry. Late Shang zooarchaeological studies depict a scenario where urban centres were supplied with lamb from nearby villages (Li 2011). Taking the large amount of sheep remains along with the absence of very young individuals into consideration, it is plausible to envisage a network where systematic trading connections had been established between urbanised centres and villages devoted to a specialised economy, such as sheep husbandry in this case.

*3.3.6 Horse (*Equus ferus caballus*)*

The mainstream of this scholarship agrees that domestic horses were introduced into China from the Eurasian steppe during the late half of the second millennium BC (Linduff 2003; Yuan & Flad 2006; Flad *et al.* 2007). While genetic evidence support the introduction of the domestic horse against indigenous origins, the possibility of gene exchange between domestic population and local wild herds has not been dismissed (Cai *et al.* 2009; Lei *et al.* 2009).

The archaeological evidence remains too vague to support either hypothesis. The presence of the wild indigenous species *Equus przewalskii* suffices to confirm human exploitation (Huang 2003; Linduff 2003). Nevertheless, as a result of the morphological similarity between wild and domestic populations, we cannot straightforwardly determine a status as wild, tamed or domesticated, especially for the initial stage of domestication (Olsen 1988; Olsen 2006).

Some Longshan discoveries of horse remains seem to be compatible with a hypothesis advocating for the introduction of horses from the steppe. These discoveries, however, were reported at a fairly early date, and the horse bones not directly dated. Some centuries later, during the late Shang dynasty, horse remains were much more frequently found. While they did occur in household contexts indicating consumption, discoveries were largely restricted to northwest China. In the Central Plains, occurrence was tightly associated with ritual offering in mortuary context. What is more, horses did not appear alone in ritual contexts. Almost always horses and chariots were buried within one another as a 'compound entity' (Piggott 1974; Campbell 2015). They together might have been perceived as a vehicle of conquest and their inhumation a mark of Shang authority. It has not been resolved yet whether these horse burials were locally husbanded or brought in as tributes to the kings (Shih 1953).

There is a surprising lack of information pertaining to horse husbandry strategy. The lack of information probably suggests that horses were of little economic utility other than being ritually significant. Many archaeologists cast doubt over this speculation and are keen on searching for horse riding evidence. The most recent study presents pathological evidence for horse-riding in Xinjiang, suggesting the adoption of this equestrian techniques by pastoral communities by the fourth century BC (Li *et al.* 2020). The pace and process of the introduction to central China remains unclear.

3.4 Animal images in artifacts

Images of animals appear in material cultures in the form of animal figurine, zoomorphic pattern, rock art and so forth. Rock art was seldom found in the Central Plains, and distribution was seemingly concentrated in either the near-steppe Ordos Region or Pacific Rim coast.

Animal figurines have been found in many archaeological sites. Most of them were made of clay during the Neolithic, and bronze during the Shang. Other materials were either perishable (*e.g.* wood) or belonged to later inventions (*e.g.* porcelain). Many Neolithic sites yield animal clay figurines – Peiligang in Henan, and Xihe and Zhangqiu in the lower reaches of the Yellow River in Shandong Province (Wang 2005). Even further from the Central Plains is the Jianghan Plains where the Yangtze and Han rivers converge and where sites belonging to the Shijiahe culture of the late Neolithic are characterised by the discovery of large collections of animal figurines. At one site, Dengjiawa, thousands of animal figurines, depicting sheep, elephant, dog, chicken, duck, tortoise and fish have been discovered (Shihe Kaogu Dui 1990). Discoveries of animal figurines have not been that numerous in the Central Plains. Occasionally animal figurines are reported, but never on a large scale. Wang (2005) proposes that most figurines represent images of domestic animals such as pigs, dogs, sheep, and cattle. The conclusion is based on a very rough estimation from selected sites regardless of the ambiguity in identifying animal taxa. A quadruped figurine from Shijiahe has even been identified as a kangaroo (*ibid.*), the animal which is endemic to the Australia. The figurine identification is far-fetched.

Zoomorphic patterns can readily be found in sites across China throughout the Neolithic and Bronze Age, depicted on a variety of materials including pottery and bronze. As it covers a wide range of animals and lacks any obvious pattern, a comprehensive review on this topic is still absent.

More often than not, zoomorphic patterns are very schematic, presenting considerable challenges towards identifying a prototype in reality. Those that are potentially identifiable are fish and birds in most cases. Further animal taxa identified on bronze artifacts during the late Shang dynasty include elephant, tiger, ox, horse, bear, sheep and boar (Chang 1981).

Admittedly this review is very brief as animal-related crafts are not the core of this study and may require a different methodology to address the problem. The purpose of this section is to remind scholars that in addition to animals on the landscape, animal images created on the mindscape are also informative for decoding the animal categories of the past.

3.5 Beastly question and domesticated scholarship: a summary

The tite of Naomi Sykes's 2015 book, *Beastly Questions: Animal Answers to Archaeological Issues,* is immediately striking. Zooarchaeology, dealing with animals, does indeed pose 'beastly' questions. In sharp contrast, most scholars are keen to explore the domestic side of the question. Such scholarship, at least in the field of Chinese zooarchaeology, has in fact itself been domesticated by the currents of the archaeological paradigm likely rooted in the modern epistemology underpinning 'scientific approach', a Darwinian theory of evolution highlighting 'domestication' and the formation of nation states possessed by the nostalgia to the remote past.

In the foregoing sections I have briefly presented the archaeological understanding of the present research period and region. Six major animals were broadly introduced along with their relationship with human beings. Two major biases were recognised in the scholarship of zooarchaeological study in China and as possibly profoundly distort our understanding of the human-animal relationship in the past.

By reviewing those studies mentioned above, the striking unbalance between the emphasis put on domestic animals and wild animals was quickly recognised. Even after agricultural lifeways were adopted during the Neolithic, the exploitation of wild animals never ceased. The late Neolithic, rather, saw an resurgence of wild animal exploitation in some regions (*e.g.* Jiangzhai and Xinzhai) and in the late Shang dynasty a diversity of wild animals were excavated from the Great settlement of Anyang (De Chardin & Young 1936). While most studies privilege the priority of domestication (*i.e.* earliest date and location when and where the domestication process occurred), little effort has been made to explore their wild counterpart. As a consequence, knowledge about wild animals along with their relationship with human beings was pales next to our understanding of domestic creatures. Lacking information about animals other than livestock presents even graver difficulties for researchers to critically review the idea of a wild/domestic dichotomy that is very possibly a modern construct. Our understanding of the past thus is held captive by modern taxonomies.

The topic of domestication itself is worthy of exploration. So is the differentiation between domestic and wild animals. The notion of 'domestic' category needs to be clarified as to its biological and social connotations: whether it refers to a category in modern biological taxonomy or it denotes a local category defined by indigenous society. The former is useful to address domestication as human action to exploit animals, while the latter casts light on human-animal relationship reflected by their perceptions of animals.

Secondly, within the subject of animal domestication, most studies aim to find the exact date and place of the earliest domestication, the landmark signaling the initial change. At the same time, limited efforts have been made to detail husbandry strategies, showing how animals were raised and interacted with humans. Although the preservation and the resolution of archaeological data may impose restrictions upon this issue, recently developed techniques such as stable isotope analysis open a new venue for investigation. What is more, zooarchaeological studies tend to be increasingly theoretically sophisticated and human-animal relationships are no longer forced to fit into stereotypical patterns – the diversity within which is acknowledged. The pig is a typical animal for which studies on raising strategies have been attempted worldwide (*e.g.* Albarella *et al.* 2006; Hamilton *et al.* 2009). There is also increasing focus upon sheep/goat and cattle management in China (*e.g.* Dai *et al.* 2014; Chen *et al.* 2015).

How animals were managed is a particularly vital issue for this study, as changes were underway in a wide range of aspects in the Central Plains from the late Longshan through the early Shang dynasty. Animal husbandry was very likely one of the areas of change. We can say the same regarding our studies of wild animals. Archaeologists know little about what was occurring 'off-site'. Nevertheless, many activities might have been conducted beyond the settlement – herding, farming, hunting, and warfare, to just name a few. Every such activity could have altered the surrounding natural environment to various degrees. The management of wildlife is one of the least understood of such issues. Gaining insight into management strategies of all animals, regardless of the latterly imposed wild/domestic division, will surely advance our understanding towards a more complicated society where humans, animals and things were tightly intertwined.

4

Animal Classification at Wadian

Our first case study is the Longshan culture Wadian site. This brief introduction at the beginning aims to render a panorama of the natural and social background at the site. We present the results of faunal analysis, from which interpretation of animal categories are inferred.

4.1 Wadian: site introduction

The Wadian site is located at Wadian Village, Huolong County Henan (**Figure 4.1**). The location is consistent with an area associated with the inception of the Xia dynasty in several ancient texts. This, in turn, corresponds in archaeological terms to Erlitou culture.

Wadian is a large-scale settlement dating to the pre-Erlitou late Longshan period. The site covers an area of 1,000,000m^2. Wadian was first discovered in 1979 during an archaeological survey conducted by the Henan Provincial Institute of Archaeology and Cultural Relics. Since that time five excavation seasons have been carried out, in 1980, 1981, 1982, 1997 and 2007 respectively. The most recent 2007 excavation season has already concluded, but no preliminary results have been published. Most data from the 1980, 1981 and 1982 excavations also remain unpublished outside of some brief reports (*e.g.* Jia *et al.* 1983). The 1997 excavation is part of the multi-disciplinary project, 'Xia-Shang-Zhou Chronology Project' to clarify the complex origin of Chinese civilisation. The 1997 season produced a thorough monograph including not only excavation data, but also providing a summary of unpublished data from earlier excavations. Our information is summarised mostly from this 1997 excavation report (HPIACR 2004), while being supplemented by other available published reports and unpublished records held at the excavating institute.

4.1.1 Chronology and phasing

The occupation of the Wadian site falls broadly into the late Longshan, which persisted for around five centuries. Chronologically, three phases can be recognised on the basis of ceramic typology. Absolute dating is derived from 14 radiocarbon dates. Phase 1 ranges from 2200~ 2100 BC, Phase 3 from around 2100~1700 BC. Absolute dates are not available for Phase 2.

Phase 1 corresponds chronologically to the late Longshan, and is characterised by the discovery of pits. Phase 2 is also characterised by pits, while urn burials and house structures have also occasionally been found. Phase 3 witnesses a dramatic increase in the number of pits and dwelling structures. Despite changes in the usage of

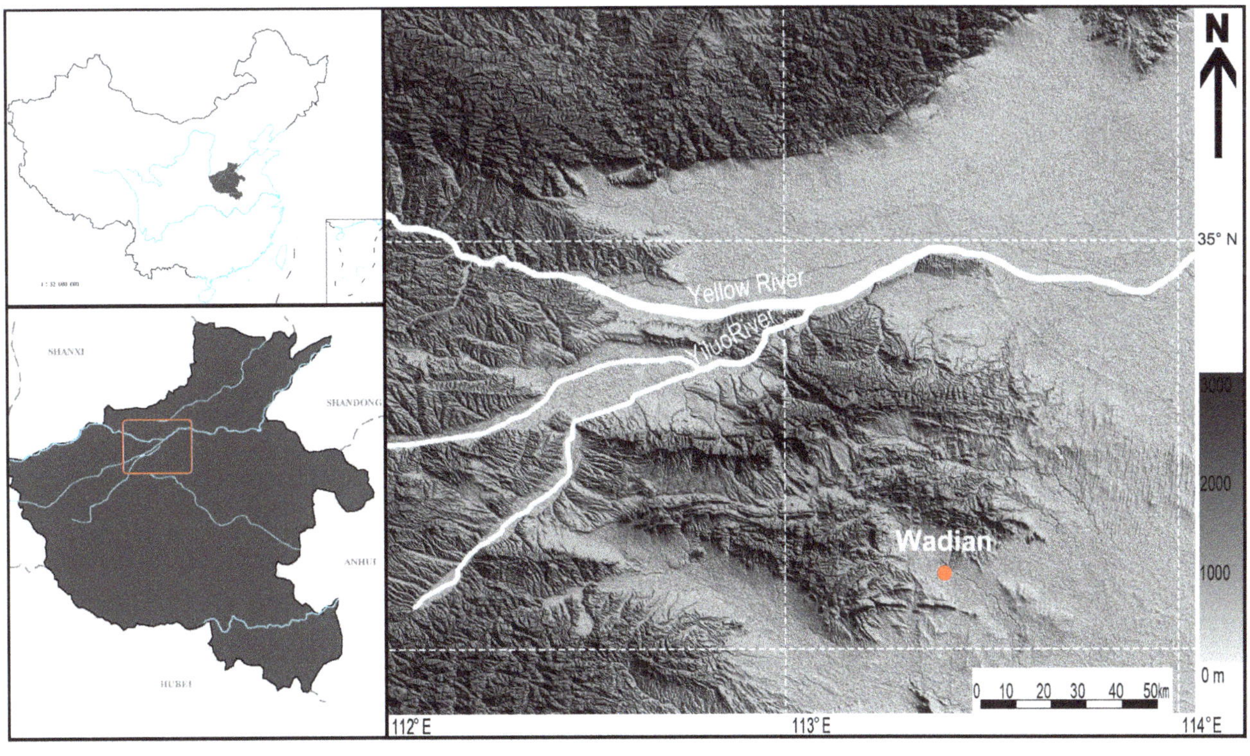

Figure 4.1 Map showing location of the Wadian site

settlement space, the ceramic typology of each phase displays consistency in pottery manufacture.

4.1.2 Spatial layout and feature types

The pre-excavation survey indicated that Wadian was a large settlement. The area excavated at present is believed to have served as the centre of the settlement. During the 1997 season excavation was comprised of two areas, covering 180 m² in total. This study only focuses on Area 1. Pits, recovered through all three phases, are the most numerous feature types discovered at the site. These features were diverse in shape, size and content.

Discoveries of large structures along with the foundation pits also typify the Wadian site. Nine house foundations together with a number of postholes and wall remnants were recovered. Five of these were semi-subterranean round houses (**Figure 4.2**). These structures were elaborately built. The ground floor of house F3, for example, was

Figure 4.2 Map of the Wadian site, showing features from three phases (modified from HPIACR 2004: figure 10)

stepped and the surface burned in order to maintain sustainability. Pedestals were made of fired clay to support stable posts. Traces of hearths are usually left in the center of houses. Wall shrines were occasionally found as well. Only a handful of burials were discovered, much fewer than expected, considering the occupation density. There were two main types of burials: three shaft burials and three urn burials. Remains of human skeletons belonging various age groups were discovered in the three shaft burials, all oriented south. No associated grave goods were found. One of the urn burials was accidentally discovered by a local farmer and associated grave goods later collected by archaeologists. Urn burial pits were ovoid in construction and one or two urns contained skeletons. There are finds of grave goods connected with these urn burials. In urn burial W1 a 25~30-year-old male was buried with a *zeng* (steamer), multiple jade pieces, and pig bones.

The number of features gradually increased from Phase 1 to Phase 3, as did their diversity. During Phase 3, as seen in the site map (**Figure 4.2**), the settlement was intensively occupied.

4.1.3 Archaeological assemblages

A large collection of fine quality ceramics was found. The drinking vessels in particular were of excellent quality. Besides from these ceramic wares, exquisitely crafted jade pieces bear witness to the artisanship of the Wadian's residents.

Archaeologists believe that Wadian provides significant clues to tracing down the origins of Xia culture (sometimes called 'Xia dynasty'). Firstly, the location of the site is consistent with the territory of the legendary Xia as mentioned in classical texts belonging to dynastic China. It is believed that archaeological evidence coincides with the description of Xia recorded in some ancient texts. For example, the grandeur of the site. Furthermore, the uneven distribution of mundane and luxurious items among different depositional contexts indicates unequal access to resources, suggesting social stratification. Animal pits found under houses are interpreted by excavators as sacrificial burials, demonstrating ritual practices and a religious attitude. Oracle bones, albeit unmarked, were recovered. Although it remains debatable whether these features explicitly match on to Xia Culture, such traits do obviously make Wadian distinct from previous sites and bear certain features that were reinforced during the dynastic periods.

(Paragraphs above summarised from HPIACR 2004)

4.2 Previous zooarchaeological studies

Lü (2009) conducted the preliminary assessment of the faunal assemblage collected during the 1997 excavation session at Wadian. A summary of his analysis (Lü 2009) follows.

Lü analysed an assemblage comprising 4829 bone fragments of bone. His research calculates both NISPs (**Table 4.1**) and MINs (**Table 4.2**) of animals. 32 taxa of animals including pig, deer, cattle, sheep/goat, dog, bird, small carnivore, tortoise, hedgehog, rabbit, fish, bird *etc.* were found at the assemblage. Amongst this assemblage pigs held the largest proportion throughout all three

Table 4.1 A summary of Lü's NISP counts and percentage of major animal taxa from Wadian

	Phase 1		Phase 2		Phase 3	
	NISP	NISP%	NISP	NISP%	NISP	NISP%
Pig	58	72.5	130	64.0	739	71.1
Deer	17	21.3	40	19.7	113	10.9
Cattle	3	3.8	4	2.0	46	4.4
Sheep/goat	0	0	14	6.9	45	4.3
Other animals	2	2.5	15	7.4	97	9.3
Total	80		203		1040	

Table 4.2 A summary of Lü's MNI counts and percentage of major animal taxa from Wadian

	Phase 1		Phase 2		Phase 3	
	MNI	MNI%	MNI	MNI%	MNI	MNI%
Pig	7	50.0	11	45.8	40	54.8
Deer	4	28.6	4	16.7	7	9.6
Cattle	1	7.1	1	4.2	3	4.1
Sheep/goat	0	0	2	8.3	4	5.5
Other animals	2	14.3	7	29.5	19	26.0
Total	14		24		73	

phases. By analysing the presence and absence of animal taxa, Lü was able to reconstruct a warm damp environment surrounded by lakes, bushes and forest. The predominance of pig bones indicates possible heavy reliance on domesticated pigs for meat consumption. Lü stressed the morphological characteristics of bones when identifying domesticated pigs from wild boar. He proposed that pigs are associated with special phenomena such as urn burials and animal pits that were built underneath house foundations probably as ritual offerings. One juvenile individual was recognised as deposited in the F8 animal pit of Phase 3. Lü provided some generalisations on the symbolic meaning of domestic pigs across prehistoric north of China and cited other examples of pigs being used as grave goods in Neolithic sites. While pig, dog and cattle remains appear from Phase 1 onwards, sheep and goat occur in Wadian following Phase 2, probably indicating artificial introduction. (Summarised from Lü 2009)

A 2012 study of animal mobility is based on the strontium (Sr) analysis of faunal remains (Zhao *et al.* 2012). The result indicates that one sheep, one cattle and two pigs beyond the local strontium range, suggesting utilisation of non-local animal resources. To explore how non-local animals were brought into the site (*e.g.* migration of living animals, trade of animal products *etc.*), the strontium study was also applied to human remains. The results indicate the existence of newcomers whose mobile pattern differed from the faunal pattern (Zhao & Fang 2014). It remains unclear precisely how these animals were brought to the site.

Chen and colleagues (2015) conducted a carbon and nitrogen (C & N) isotope study on animal diet at Wadian. The isotopic values reveal that dog and pig had a dietary intake based largely on consumption of millets or millet byproducts, while cattle, in contrast, consumed a large proportion of C3 plants including rice and wild grasses, two types of plants which were also found along with macro remains (*ibid.*). This differential access to crops and plants between animals might have been influenced by both animal behavior and human management. The mobility pattern illustrated in strontium isotope studies above again adds to complexity when deciphering the interaction between animals and human beings at Wadian.

4.3 Results

The NISP of each animal taxon has been recalculated due to the lack of a clear definition of 'identifiable' in previous studies. I have refined 'identifiable' specimens under both anatomically and taxonomically recognised bones. Details of this identification method can be found in section 2.4.1. **Table 4.3** summarises the NISP results according to this modified counting method, with percentages plotted in **Figure 4.3**.

Table 4.3 Revised NISPs of major animal taxa from Wadian

	Phase 1	Phase 2	Phase 3
Pig	57	99	713
Deer	17	30	107
Cattle	3	4	40
Sheep/goat	0	14	46
Other animals	7	73	123
Total	84	220	1029

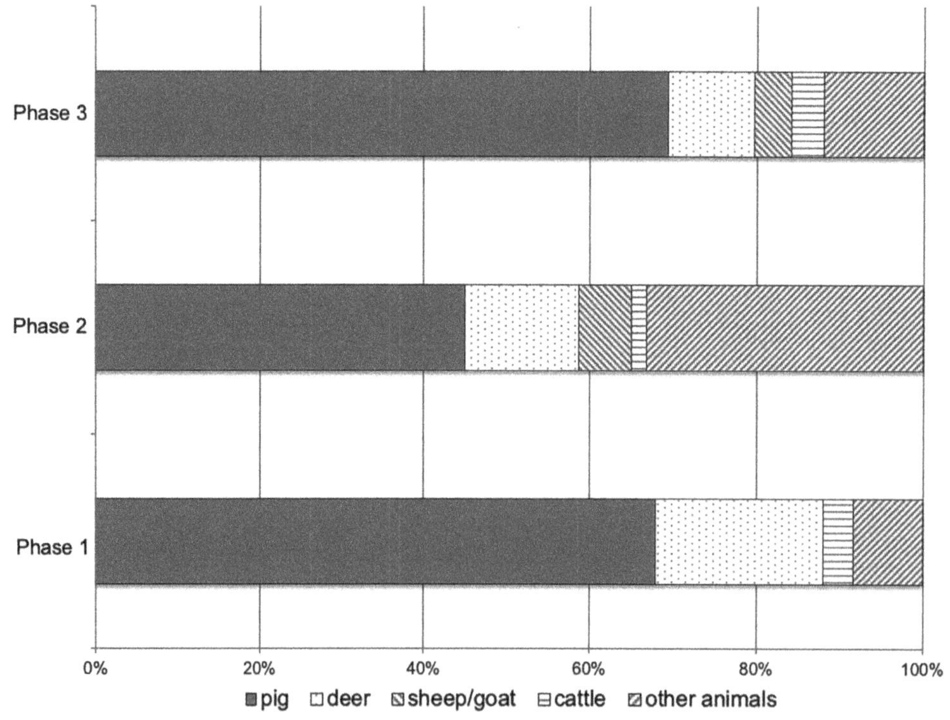

Figure 4.3 Revised NISP percentage of major animal taxa at Wadian

Four major animal taxa are compared. 'Other animals' category denotes those animals belonging outside of the four major categories, including dogs. Our overall distribution and temporal trends remain similar to Lü's result's. For consistency and clarity the following analysis uses revised NISP counts. . We focus on Area VI, given the lack of any attested connection at the distant Area V.

Calculating identifiable ratios (NISP: TNBF (Total Number of Bone Fragments)) is a straightforward method of demonstrating degrees of fragmentation. The higher the percentage, the less fragmented the accumulation. As illustrated in **Table 4.4** and **Figure 4.4**, assemblages from most contexts have identifiable ratios ranging from 10-50 per cent. This seems to represent an average range of fragmentation. All deposits from layer contexts fall within the average range and other contexts in this range include pits and houses (surface and floor underlay). Deposits with identifiable ratios of over 70 per cent are mostly found in urns (buried in H17), walls (F2-2), urn burials (W1), animal pits in houses (F2 animal pit) and pits near houses (H54 & H25), whereas sub-60 per cent ratios include more pits spatially associated with houses (H35 & H36). Contexts in the 90-100 per cent range also increase in comparison with other non-average groups. Contexts within this range also display some distinct characteristics in other aspects such as higher pig-deer indices, more associated bone groups and so forth. This is discussed further below.

4.3.1 Fragmentation

Pits were the principal feature type discovered at Wadian. Three types of pits were distinguished according to the shape of the opening: round, rectangular and irregular. Identifiable ratios were also calculated for each pit and plotted into scatter chart using different markers for opening shape. As shown in **Figure 4.5**, though identifiable ratios of pits in each group are consistent, the

Table 4.4 Identifiable ratio of assemblage in each context at Wadian

Context	Identifiable ratio (%)	Number of identified specimens	Context	Identifiable ratio (%)	Number of identified specimens
F1	21.5	17	H33	40.0	20
F2 animal pit	100.0	143	H34	24.1	7
F2-2	83.3	5	H35	60.0	3
F2 underlayment	21.9	7	H36	62.5	15
F3	22.2	14	H37	24.8	29
F4	14.5	12	H41	0	0
F5	33.3	10	H45	100.0	4
F7	28.6	4	H47	8.3	1
F8 animal pit	9.2	8	H51	25.0	24
H1	26.1	6	H53	8.8	3
H2	15.7	8	H54	100.0	3
H3	49.3	82	H58	21.9	7
H4	20.0	2	H59	19.0	4
H5	35.7	20	H61	12.6	25
H8	16.7	20	H63	50.0	2
H10	23.1	3	H64	29.0	9
H11	27.6	8	H65	12.4	13
H12	50.0	2	H66	33.3	3
H14	32.6	14	H67	30.0	6
H15	42.9	3	TI layers Phase 3	41.7	25
H17	11.7	7	T2 layers Phase 3	25.1	77
H18	20.6	14	T3 layers Phase 1	27.0	17
H19	30.8	4	T3 layers Phase 2	20.4	56
H20	27.8	20	T3 layers Phase 3	29.6	151
H24	18.7	58	T4 layers Phase 1	42.9	18
H25	100.0	1	T4 layers Phase 2	26.0	19
H27	33.3	7	T4 layers Phase 3	21.2	123
H29	38.0	19	T5 layers Phase 3	24.7	20
H30	21.6	8	T6 layers Phase 3	31.1	56

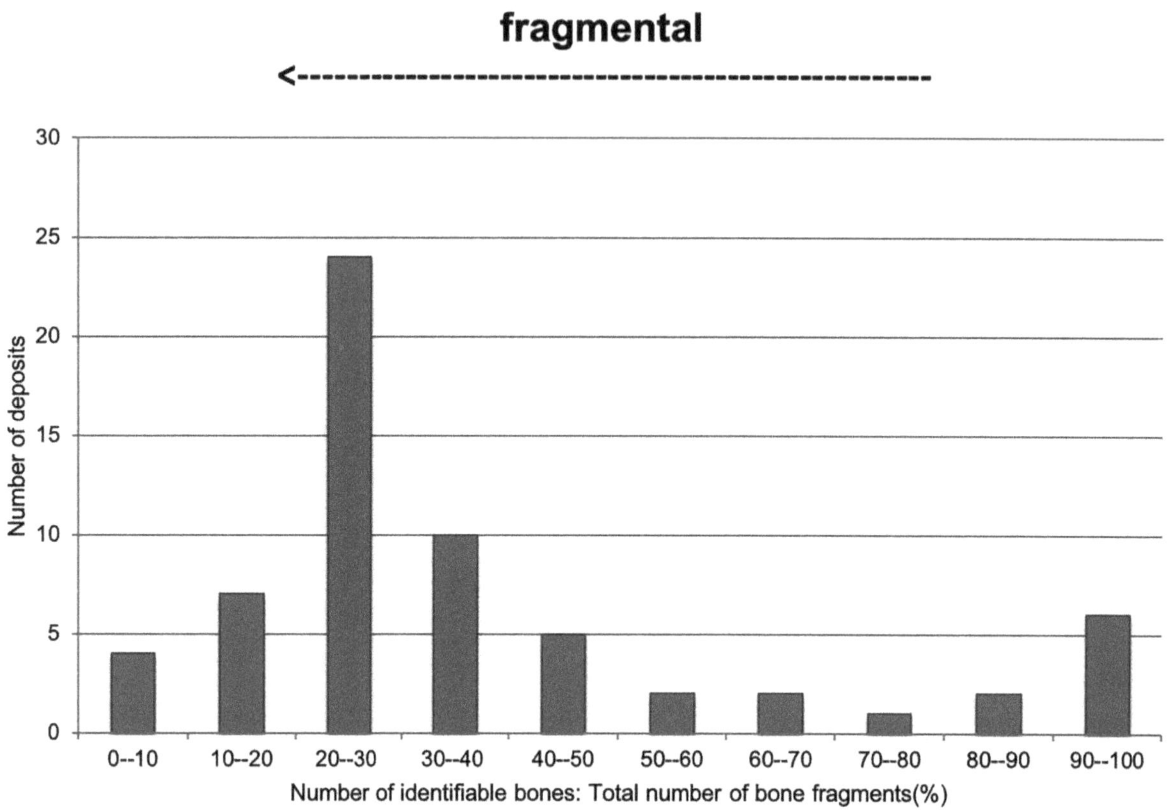

Figure 4.4 Distribution of identifiable ratios in depositions at Wadian, suggesting assemblage fragmentation

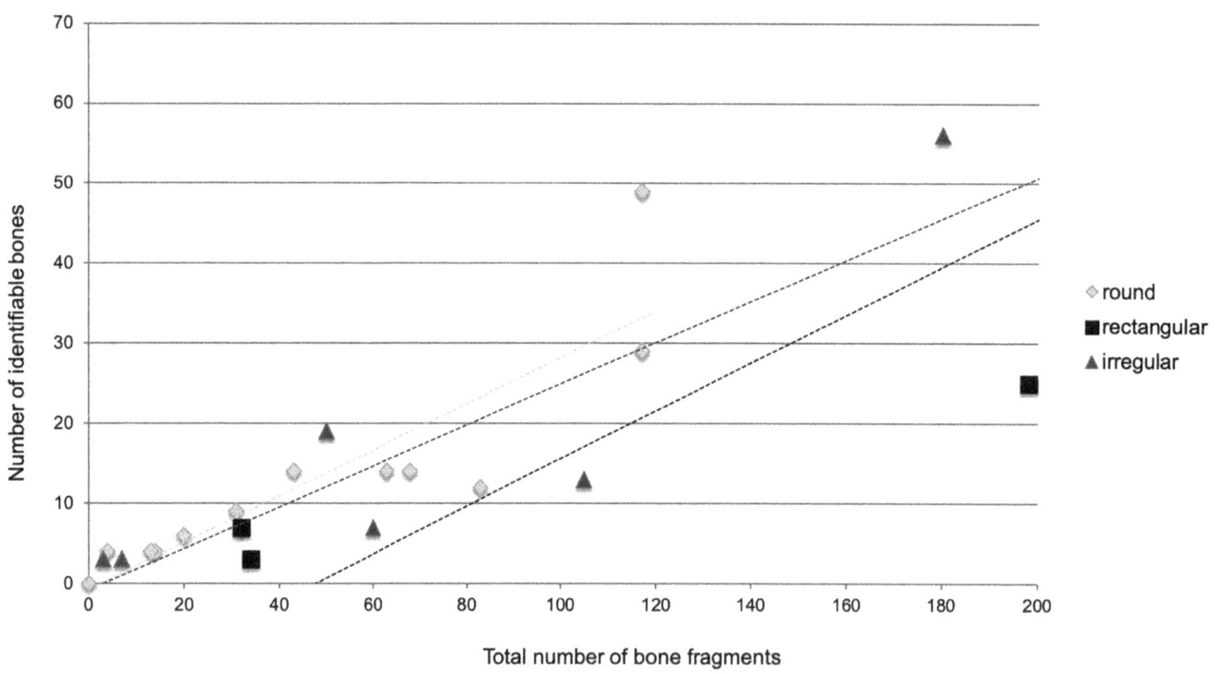

Figure 4.5 Identifiable ratios among pits with different opening shapes at Wadian

differentiation between the three types is not statistically valid when subject to a t-test ($p_{\text{round vs rectangular}} = 0.9$, $p_{\text{rectangular vs irregular}} = 0.8$, $p_{\text{round vs irregular}} = 0.5$). In other words, there is no significant statistical difference in the fragmentation of faunal depositions within different-shaped pits.

4.3.2 Body element distribution

Distributions of skeletal elements of four animal taxa (pig, cattle, sheep and deer) from each phase are summarised in **Table 4.5** and plotted in **Figure 4.6**. Specimens from feature contexts are distinguished from layer contexts.

Table 4.5 Body part representation of major animals from Wadian

		Feature contexts				Layer contexts			
		Pig	Cattle	Sheep	Deer	Pig	Cattle	Sheep	Deer
Phase 1	Head	14	0	0	1	2	0	0	3
	Axial	4	0	0	0	1	0	0	0
	Forelimb	8	0	0	2	3	0	0	2
	Hindlimb	1	0	0	2	3	0	0	1
	Extremity	2	0	0	2	2	0	0	2
Phase 2	Head	22	1	1	4	19	0	1	3
	Axial	4	0	0	2	3	0	1	0
	Forelimb	15	0	3	4	13	0	0	1
	Hindlimb	10	0	0	2	8	0	1	2
	Extremity	6	2	4	7	5	0	3	4
Phase 3	Head	78	1	3	9	97	1	1	7
	Axial	7	4	2	0	13	2	0	3
	Forelimb	45	1	12	14	62	3	3	14
	Hindlimb	22	6	4	6	26	3	1	4
	Extremity	28	5	4	10	27	6	20	16

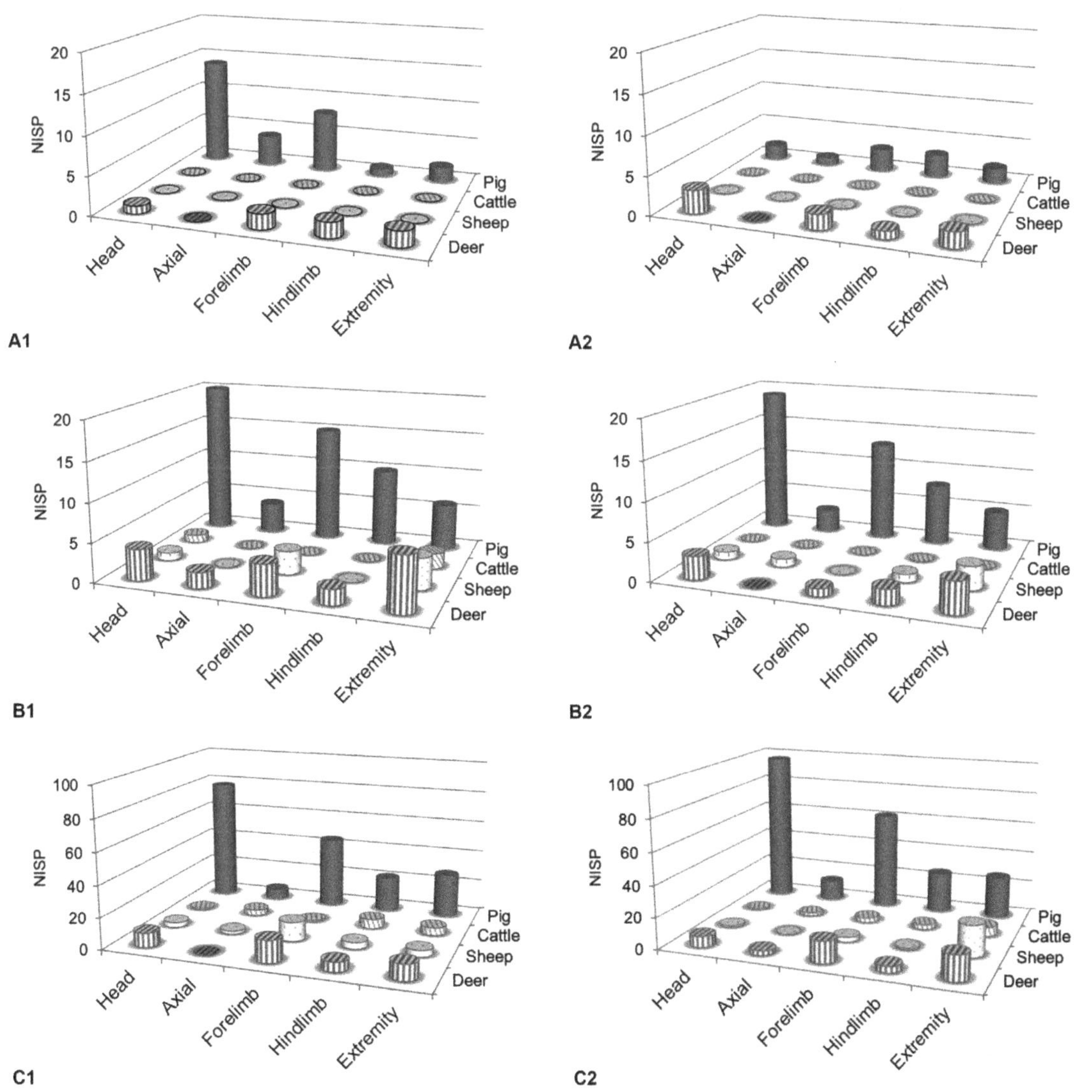

Figure 4.6 Body part representation of animals at Wadian in: (A1) Phase 1 feature contexts, (A2) Phase 1 layer contexts, (B1) Phase 2 feature contexts, (B2) Phase layer contexts, (C1) Phase 3 feature contexts, (C2) Phase 3 layer contexts.

In Phase 1, the distribution patterns of body elements in both pig and deer demonstrate slight differences between feature and layer contexts. While head elements followed by forelimbs rank as the most substantial part of the assemblage from feature contexts the proportion of head elements drops in layer contexts, and limbs predominate instead. Deer head specimens outnumber other body parts in layer contexts but drop to the bottom in feature contexts. A limited sample size is perhaps one reason for this slight departure from other phases. A paired Wilcoxon test was run in both feature and layer contexts to test the difference of body part distribution pattern. The results suggest that certain degree of resemblance between deer and pigs cannot be dismissed (p=0.29 in feature context, p=0.53 in layer context). No data is available for sheep and cattle.

During Phase 2 body part representation from feature and layer contexts resemble each other. The head is the most numerous body part for the pig in both feature and layer contexts, and is followed by the forelimb, hindlimb, extremity and axial elements. Pieces from the extremities are the most numerous part in feature and layer contexts in the case of deer and sheep. Axial elements are relatively smaller in number and proportion for both deer and sheep. Cattle bones are absent in layer contexts whereas in feature contexts most cattle bones belong to the extremities. Differences in body part representation among four animal taxa have been examined statistically. Body part representations of sheep and cattle share a certain degree of resemblance in feature contexts (p=0.60). Similarity between the remaining animal groups is not statistically significant. In layer contexts, deer and sheep display the closest resemblance in body element distribution (p=0.33) when compared to other animals (p=0.09). Due to the absence of cattle bones, no comparison of cattle was conducted.

In Phase 3, head specimens were most numerous for pigs, followed by forelimb, extremity, hindlimb and axial elements. The pattern was similar for both feature and layer contexts. For deer in feature contexts, forelimbs rank first, closely followed by elements from the extremities; while in layer contexts bones from the extremities outnumber forelimb elements, though the absolute numerical difference is small. Cattle bones are mainly from the hindlimbs and extremities, for both feature and layer contexts. Sheep remains are concentrated in forelimb component in feature contexts, whereas in layer contexts most sheep bones belong to the extremities. Statistical results show a striking resemblance between the element distributions of sheep and cattle (p=1). The body part representation of pig is different for all other animals ($p_{\text{pig vs sheep}} = p_{\text{pig vs cattle}} = p_{\text{pig vs deer}} = 0.09$). Deer and sheep, deer and cattle share certain similarity in the patterning of body elements ($p_{\text{deer vs sheep}} = 0.27$; $p_{\text{deer vs cattle}} = 0.29$). In layer contexts, the body element distributions in sheep and cattle are almost identical (p=1). Body elements from deer and sheep distribute in a relatively similar pattern (p=0.33) while the degree of resemblance between deer and pig, deer and cattle, pig and sheep, and pig and cattle are remarkably reduced (p=0.09). Seemingly no significant departure is observed in body element distribution between feature and layer contexts.

Chronologically, no obvious change occurs in body part representation throughout the three phases. For body part representation in pig, the head elements always constitute the largest proportion, except for the layer contexts of Phase 1. This emphasis upon the head is not found in other animals for any other phase and context. For the remaining animals, the general pattern of body part representation is represented by a concentration of limb and extremity elements.

4.3.3 Age profile

Detailed dental eruption and wear information that would permit a reconstruction of age profiles are not available to the study at present. Lü (2009) presents the age profile of pig as estimated by the mandibles. This estimate suggests that individuals in the range of 1.5~2 years old were consumed. Pig age profile is estimated on the basis of epiphyseal fusion, which adds another line of evidence. Age is estimated at a precision of roughly one year. As shown in **Table 4.6** and **Figure 4.7**, pigs were largely slaughtered in their second year throughout all three phases. The result, based upon fusion information, is consistent with an age estimation on mandibles, during Phase 3, with pigs consumed at a slightly younger age compared to the previous two phases, suggested by the increase of unfused bones belonging to the middle-fusion group.

4.3.4 Presence of human bones

Human bones were found exclusively in burial contexts. They were not mixed with animal bones in pit and layer

Table 4.6 Pig NISPs in each age groups estimated by epiphyseal fusion information at Wadian

	Early fusion		Middle fusion		Late fusion	
	Unfused	Fused	Unfused	Fused	Unfused	Fused
Phase 1	3	7	1	2	2	1
Phase 2	5	17	4	5	13	2
Phase 3	10	12	10	7	16	10
Total	18	36	15	14	51	13

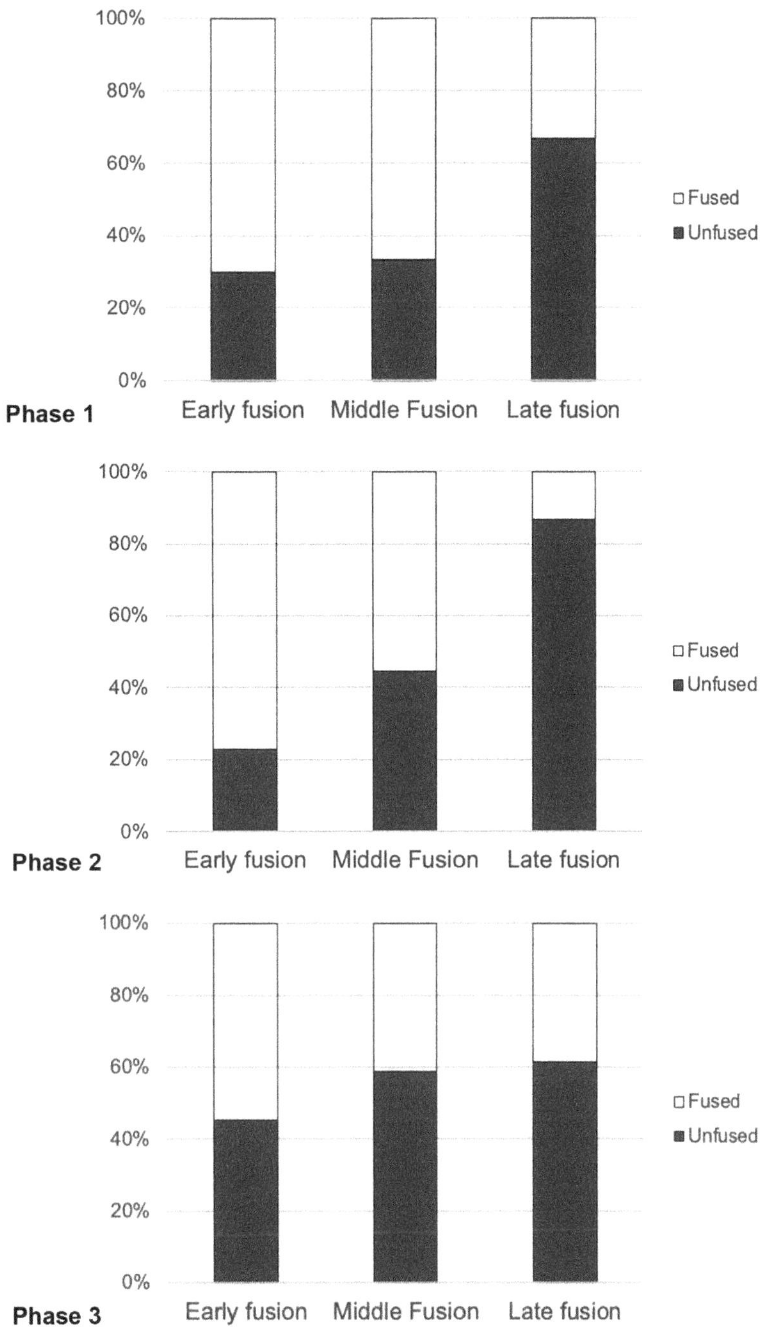

Figure 4.7 Epiphyseal fusion percentage of pigs from Wadian

contexts. In one urn burial W1, pig bones were unearthed along with a child skeleton.

4.3.5 Pig-deer index

Pig-deer indices (p-d indices) were calculated to explore the potential differentiation in deposits of pigs and deer respectively. When p-d indices are plotted into boxplot, a temporal shift in ranges of indices is clear (**Figure 4.8**). A great expansion of range into the pig-rich end is observed from Phase 1 to Phase 2. It was also during phase 2 that deposits of pig bones exclusively (index=1) appear. The central tendency, as illustrated by the medians, however, remains steady throughout the three phases.

To further explore this differentiation in spatial layout, the distribution of indices is visualised on the site map in shades of grey. As mentioned in section 2.4.3, increasingly dark grey denotes depositions rich of pig remains (index≥0.6), while increasingly lighter grey indicates deer-rich depositions (index≤0). The remaining depositions are shaded in intermediate grey, indicating neither pig-rich nor deer-rich contents.

In Phase 1, light and moderate greys are randomly scattered in the south corner, and dark grey is absent (**Figure 4.9**). During Phase 2, dark greys, indicating exclusive deposits of pig bones, appears in the urn burial W1 (**Figure 4.10**). In Phase 3 when two large-scaled longhouses surrounded

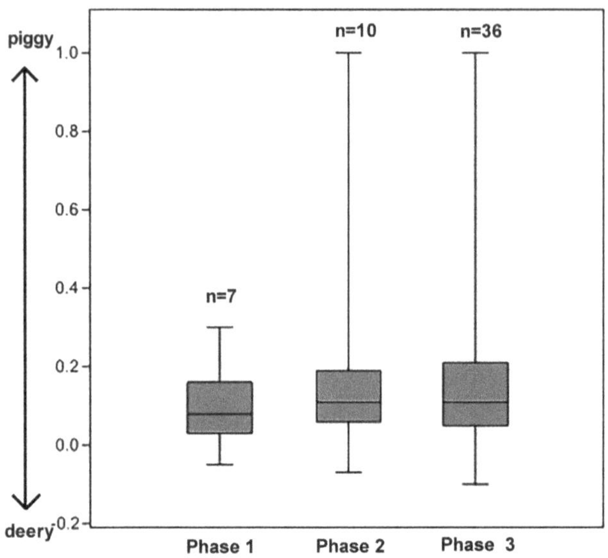

Figure 4.8 Boxplot of pig-deer indices (Note: F6 of Phase 1 is excluded as the feature is only partially excavated and its boundary remains unclear)

Figure 4.9 Spatial distribution of pig-deer index in Phase 1 at Wadian

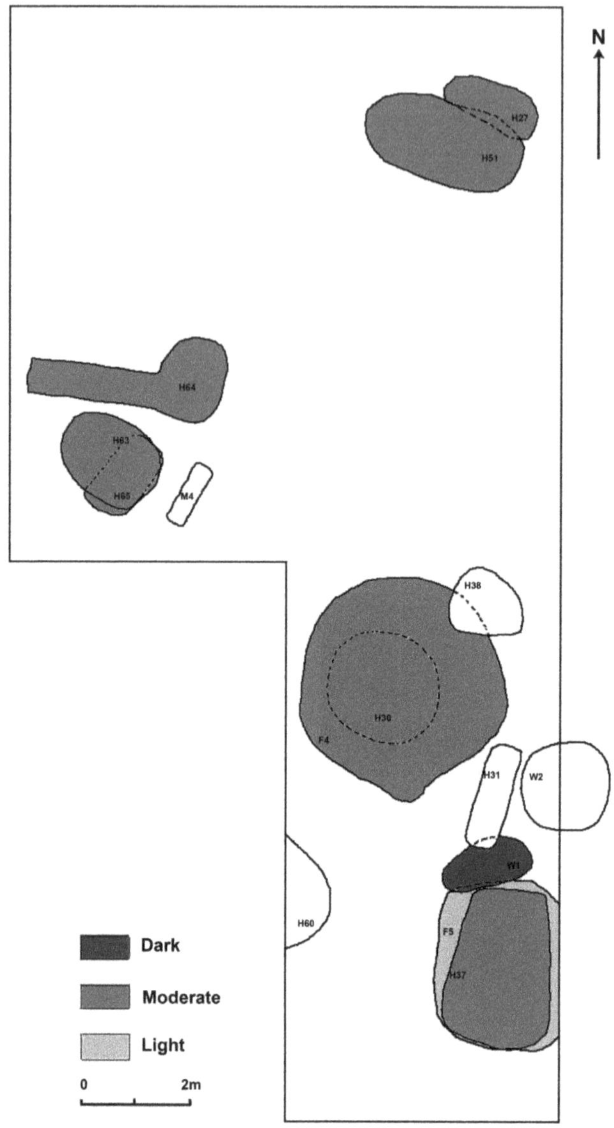

Figure 4.10 Spatial distribution of pig-deer index in Phase 2 at Wadian

by a number of burials and pits were found in the north, dark greys appear in the adjacent of or within these two houses (**Figure 4.11**).

4.3.6 Associated bone group

Most associated bone groups (ABGs) were recovered in layer contexts dating to Phase 2 and Phase 3. ABGs are present in five pits: H3, H33, H24, F2 animal pit in Phase 3 and H65 in Phase 2. The ABGs found in animal pit F2 are composed of 24 fragments from a juvenile pig. Additional traits of these pits will be discussed shortly. Of these pits, H65 has a rectangular opening while the remaining three have round openings.

Most ABGs (8 out of 13 groups) come from pig. There are, additionally, three groups of deer ABGs, one group of sheep/goat ABG and one group of cattle ABG. The number of bone specimens in each ABG is always less than five, except for Phase 3 layers in T3 (NISP of ABGs= 54) and F2 animal pit (NISP of ABG= 24).

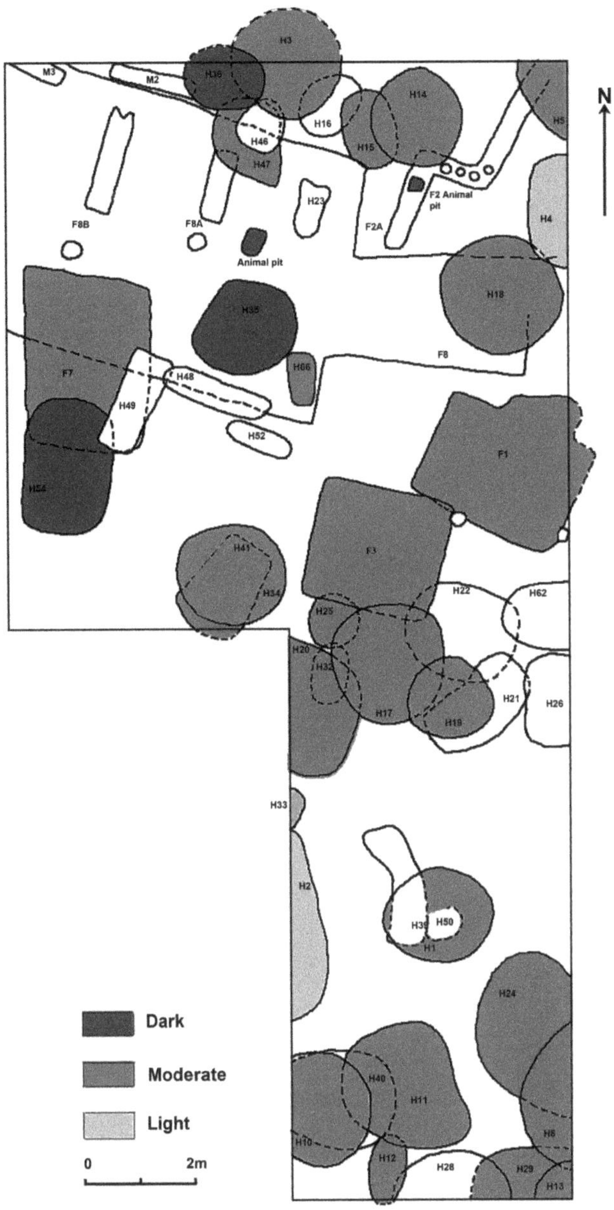

Figure 4.11 Spatial distribution of pig-deer index in Phase 3 at Wadian

4.3.7 Scatter plot of combination of two variables

Not every combination of variables produced meaningful results. However, comparisons between associated bone groups and identifiable ratios, pig-deer indices and deer NISPs do reveal certain intriguing patterns.

In **Figure 4.12**, the identifiable ratio is compared with the number of ABGs for each context. Those deposits presenting ABGs are clustered within the average range of fragmentation. These are pit H65 and T4 layer from Phase 2 and pit H24, H3, H33, T6 layer, T4 layer, T3 layer and T4 layer Phase 3. F2 animal pit is the only context with ABG beyond the average fragmentation range.

F2 animal pit is also distinct from others, as shown in **Figure 4.13**, in which number of ABG contexts is plotted against pig-deer index of each context. The dot distribution resembles **Figure 4.12**, where the data point of the animal pit F2 lies beyond the main cluster on the very pig-rich end. All contexts presenting ABGs are located on the pig-rich side (>0) in the chart. Compared with **Figure 4.14**, where data points of contexts with ABGs span over the whole range of deer NISP, the possibility of distortion in distribution due to deer sample size can be tentatively eliminated. Thus F2 animal pit is in general distinct from other depositions regarding its identifiable ratio, pig-deer index and associated bone group deposit.

4.4 Discussion

As discussed in chapter two, every classification scheme is strongly contextual. It therefore becomes vital to understand the context in which knowledge about animal categories was achieved and applied in practice. Before investigating the domain of animal classification at Wadian, I begin by exploring the site itself, attempting to grasp the background that was set for humans' perception of the natural and social world there.

4.4.1 Area themes

Wadian features three phases, recognised as spanning over 500 years (HPIACR 2004). Although each feature is assigned to one of these phases based on stratigraphy, contemporaneity between features cannot be presumed. Data resolution in fact presents difficulties to an advanced exploration of the issue. However, a pattern of some consistency remains, what I would call an 'area theme'. The meaning of 'theme' is twofold. Firstly, it expresses how features bounded within this particular area share certain characteristics. Such a characteristic may refer to many things – namely similarity, functional complementarity, symbolic resemblance and so on. Secondly, overall area characteristics were fixed and inherited across generations. Intra-area rearrangement and reconstruction therefore was always permitted, provided shared characteristics of features within this area remained consistent. The use of 'theme' thus takes the discussion away from ambiguity relating to contemporaneity and toward a positive maintained anthropic structure.

It seems that during Phase 1 and 2 the theme of each area might have not yet formed. In Phase 1, a single house along with other several pits was found in the southern part of the excavation area (trenches T3 & T4) alone (**Figure 4.9**). There may have been more features to the south of the excavation area. However, no evidence has been collected so far. An expansion of settlement occupation towards the north in Phase 2 is demonstrated in **Figure 4.10**. During Phase 2 population began to expand to the north with burials and pits appearing in the central and northern part.

It was in Phase 3 that three distinct areas emerged (**Figure 4.11**). Pits feature the southern part (trench T4), where earliest Phase 1 remains were clustered.

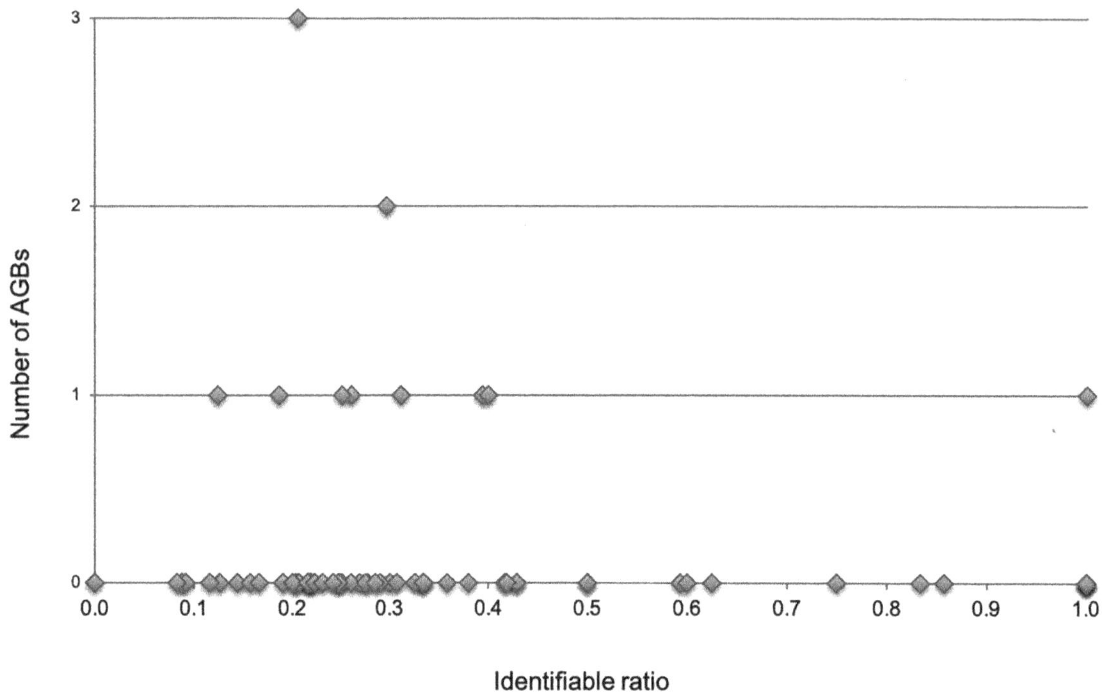

Figure 4.12 Scatter plot of number of ABGs against identifiable ratio in each context at Wadian

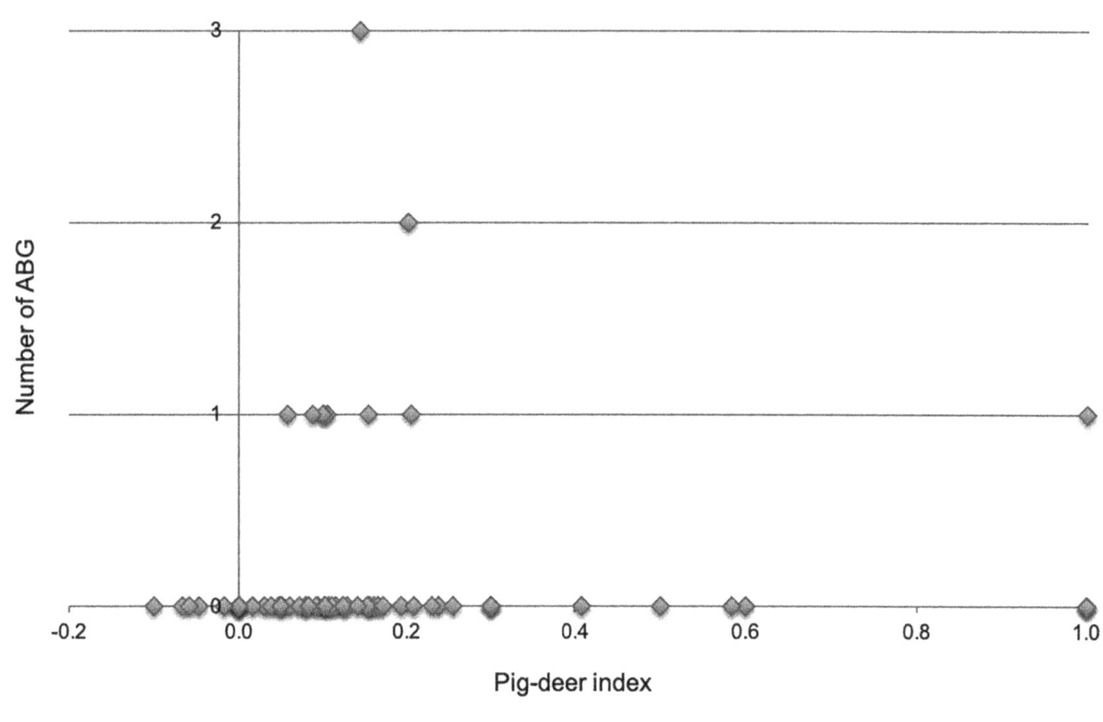

Figure 4.13 Scatter plot of number of ABGs against pig-deer index in each context at Wadian

The theme of the central site component, roughly covering trenches T2, T3 and T6, is a trend towards semi-subterranean houses and pits. Striking resemblances can be observed between houses F1 and F3 with regard to size, shape, and orientation.

Two sizeable long-houses surrounded by several burials and pits form the theme of the northern part (trenches T1 & T5). Houses and pits which cut into each other, indicating succession in sequence. In the long-house F2, wall F2B cut into wall F2B, which is suggestive of restoration or reconstruction events. Therefore at least two episodes featuring of changes in spatial layout can be tentatively identified. Both houses F2 and F8 were constructed stratigraphically beneath Layer 3. House F2 seemed to cut into the northern edge and two internal walls

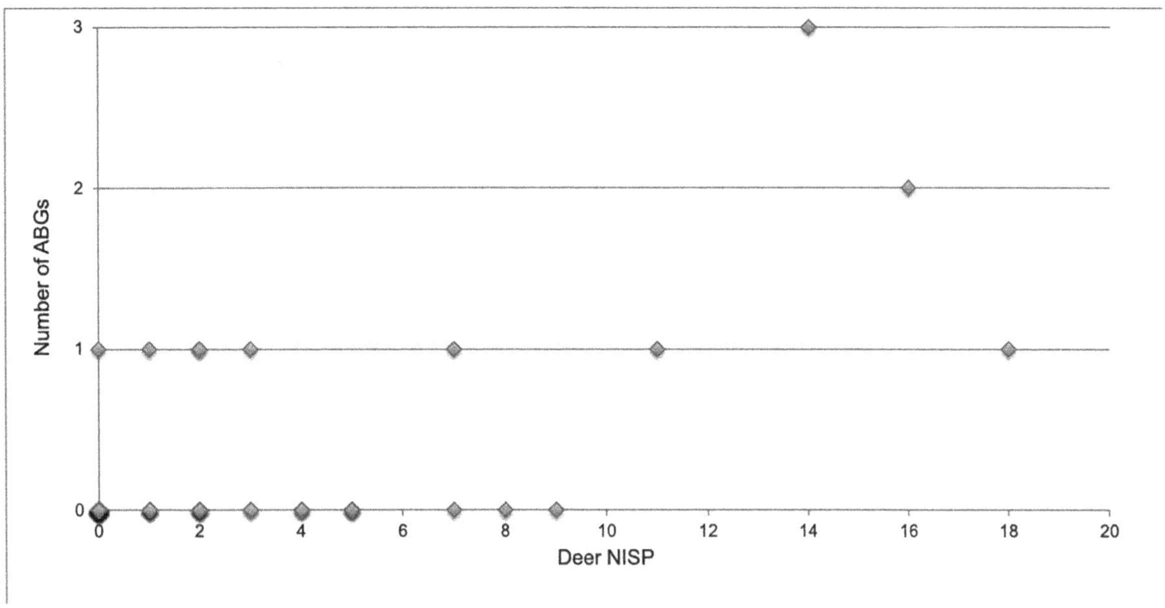

Figure 4.14 Scatter plot of number of ABGs against deer NISP in each context at Wadian

of the foundation of house F8 F8A & F8B. Most likely the architects of longhouse F2 were fully aware of the structure and location of F8. The construction of house F2 could have taken place immediately after the abandonment of house F8 when the outline and inner structure of house F8 were both still intact. Regarding this vague boundary in space and time, it was very likely that the reconstruction process was continuous and the consistency in area theme was carefully maintained.

The theme was not of just a single phase, but maintained throughout the history of the sit, as manifested by the consistent differentiation in using two types of houses. Houses at Wadian fall into two categories: the sizeable ground-level houses including house F6 of Phase 1, house F2 and F8 of Phase 3, and the small-scaled semi-subterranean houses such as F4, F5 of Phase 2 and F1, F3, and F7 of Phase 3. If both types of houses served different purposes either in functional or symbolical sense, a broader variety would be observed in other respects as well.

Internal partitions were found exclusively in longhouses. There were wall remnants in both house F8 and house F2. Nevertheless they were not necessarily contemporaneous. In the case of house F2 especially the wall F2B is reported as having cut into the wall F2A. House F6 and pit H67 in Phase 1 are recorded as two individual features, although they were very likely originally components of a single house. Pit H67 was connected to house F6 by two wide steps and the composition of their layers resembles each other stratigraphically. In addition, traces of a hearth were left in house H67, which comprises a house unit ideally. It is unclear if the space could have been further divided depending on house size. Alternatively, divisions were constructed according to different functions.

It was rare to find faunal remains on the surface of large-scale houses such as houses F6 in Phase 1, and F8 in Phase 3. The longhouse F2, however, was covered by a layer of plaster floor which contained several bone fragments.

Animal pits beneath foundations are characteristic of longhouses. In the case of house F2, an entire juvenile pig was buried under the wall F2A. A rectangular pit was dug into the house foundation of F8, and contained more than 80 fragments of largely pig bone. The original excavation report tends to link such animal pits with ritual ceremonies associated with house constructions (HPIACR 2004). A discussion on animal sacrifices and ritual ceremonies will be presented in later sections.

Both types of house show evidence of abandonment and reuse. At the longhouses, several layers of floors were accumulated in succession in house F6. There were at least two distinct episodes of occupation, indicated by a layer of grey soil which separated compactly constructed floors (HPIACR 2004). This is less certain in the case of house F8 where two walls, instead of overlapping, stood only 1.5m apart (HPIACR 2004). As this interstitial space seems too small to accommodate even a single person, it remains unclear whether the walls were constructed to make a room for specific use or otherwise were simply not contemporaneous. Occasional traces indicating reuse of semi-subterranean houses were found. Semi-subterranean house F5 is the only case where layers of accumulation were identified. We should mention that house F5 lay directly beneath the massive pit H37, which was similar in size to house F5 (*ibid.*). The opening of pit H37 was in a layer that postdated F5, suggesting that the house preceded the pit. An assemblage of ceramic containers and several bone artifacts were recovered from this opening (*ibid.*). The function of this pit and its relationship to house F5 await investigation. The evidence so far seems to support the speculation that restoration and/or reconstruction were probably limited to large-scale ground houses.

Ceramic fragments were found in every house and no great difference in pottery type has been discovered among houses. Apart from pottery, bone and lithic tools including stone spindle whorls, lithic chisels, bone hair pins, bone awls, needles and others were also recovered in smaller houses such as F1, F3, F7 in Phase 3 and F5, F4 in Phase 2 (*ibid.*). It is hard to categorise these artifacts and thus further assert connections to labour division and domestic production.

Burials were placed only within the longhouses such as F8 and F2 in Phase 3. They were neither within semi-subterranean houses nor closer to them. The urn burial W2 is the sole exception. Being cut into the edge of house F5, it was unlikely contemporaneous to that feature. W2 contained a set of delicate jade artifacts including jade bird and spade. An adult male was buried in the urn, together with pig remains.

In short, the differentiation between houses and their occupation patterns were consistent throughout the three phases, which resonates with the stability of area themes. A strong tradition was passed down from generation to generation of inherited patterns in both area theme and house type. Hodder & Cessford (2004) advocate using the term 'social memory' in lieu of 'tradition' so as to emphasise the active role of the human in constructing and maintaining such social norms during their daily life. This is archaeologically apparent at Wadian. Such consistency of pattern might have been even stronger at Wadian when taking isotope results into account. It has been found that Sr isotope ratios from several human and animal (including cattle and pig) bones fall out of local average range, which suggests that people and animals of non-local origins once lived in Wadian (Zhao & Fang 2014). Instead of creating the diversity in material representation, residents of various backgrounds adapted to the local institutional structure and even positively participated in its maintenance, as suggested by the stability of area themes and house types. A form of strong centripetal power might be speculated, which would have ensured the cultural continuity manifested in materials.

4.4.2 Site formation processes

The results presented above illuminate several types of deposition formed by different human depositional actions. Broadly speaking, five types of depositions were recognised based on a series of characteristics: foundation depositions, house-related pit depositions, house floor depositions, pit depositions and depositions in archaeological layers outside feature contexts.

Table 4.7 compares different types of depositions and summarises their features. Among them foundation depositions and house related pits could be roughly grouped together when we consider their house-related spatial distribution and pig-rich infills as especially indicated by percentage of pig remains in one deposition and the number of such pig-rich depositions. In terms of fragmentation, foundation depositions display a moderate identifiable ratio probably due to the lower ratio in F8 animal pit owing to the large number of unidentifiable vertebrates and ribs in the assemblage. Therefore fragmentation ratios in foundation depositions are very likely over-estimated. In contrast, house floor and pit show lower identifiable ratios along with lower pig percentages and juvenile pig deposits are absent in these contexts. Pigs regardless whether they were mature or juvenile, wild or domestic were placed in mixed deposits within pit and layer contexts. As previously stated, structured depositions were the result of intentional human practice. As inferred from the evidence above, floor and pit depositions were likely results of refuse accumulations while foundation

Table 4.7 Types of depositions at Wadian

Deposition type	Foundation	House-related pit	House floor	Pit	Layers
Context	F2 & F8 animal pits	H3, H4, H5, H14, H15, H17, H18, H25, H35, H36, H47, H54 & H66,	F1, F2-2, F2, F3 & F7	H1, H2, H8, H10, H11, H12, H19, H20, H24, H29, H33, H34	
Fragmentation (identifiable ratio in %)	54.6 (100; 9.2)	49.3	40.7	28.7	34.1
Number of ABGs	1	1	0	4	8
Average pig percentage (%)	100	58.0	44.0	57	40.6
Number of pig-rich deposits ($NISP_{pig}$ / $NISP_{total} \geq 90\%$)	2(n=2)	5(n=14)	0 (n=5)	0 (n=12)	0 (n=10)
Presence of juvenile pig deposit	Yes	No	No	No	No
Presence of wild boar deposit	No	No	No	No	No
Mixture of pig deposit	No	Yes	Yes	Yes	Yes

and house depositions suggest certain purposes other than refuse dumping. Put together, we can speculate that depositions indicate two kinds of formation processes, the former associated with daily refuse disposal and the latter with special events serving particular purposes.

4.4.3 Temporal change of category

As seen in both boxplot and index map (**Figures 4.8 & 4.9**), during the earliest phase at Wadian, animal bones were deposited together, regardless of their taxonomic categories. During Phase 2, pig-rich depositions appeared in the form of urn burial (W1) (**Figure 4.10**). Pig remains were found placed in an urn burial accompanying an adult male skeleton. This urn burial W1 is singled out from other burials in that delicate jade objects were recovered as grave goods and the burial was dug abutting the north wall of the house F5 (HPIACR 2004). During Phase 3, a distinct pattern concerning depositing pigs and deer emerged in which the overwhelming body of pig-rich deposits was found in the area under the theme of longhouse (**Figure 4.11**). This temporal change in the patterning of bone depositions might have reflected changes in animal categories throughout the time. It is speculated that the distinction between pigs and deer was gradually established through Phase 1 to Phase 3 and that categories were reinforced during Phase 3, and was reflected in the patterning of archaeological remains.

The predominately mixed depositions of Phase 1 does not provide support for a clear division between animals. However, it does not mean that there was no categorisation of animals at all. But we may plausibly speculate that such a classification, if any, might have not been widely practiced in daily life. Consequently it left no trace in archaeological assemblages. It is also worth noting that Phase 1 pits made up the majority of the feature types, most of which were likely refusal depositions. It might not have been necessary to deliberately separate animals when disposing them in refuse dumps.

The appearance of pig-rich depositions associated with houses during Phase 3 indicates a separation between animal categories. Classification of animals is first and foremost a type of knowledge shared by a community. The reinforcement of animal categories has a two-fold significance. Firstly, contents became increasingly steady and durable. In other words, boundaries between any two taxonomic categories emerged and members in each taxonomic category experienced little change. Secondly, this knowledge was coded into social rules that directed various activities within the community. Archaeological assemblages are averaged out by time and space (Bailey 2007). The accumulation of pig bones in pits in the vicinity of houses was the result of behavioral repetition. It was very likely that certain social tenant might have underpinned this consistent utility pattern displayed in Phase 3.

The occurrence of mixed depositions in majority of instances, with one example of pig-rich urn burial in Phase 2 seems to suggest that Phase 2 was probably a transitional period.

The emergence of sole pig deposits reflects that 'pig' as animal category had formed a category distinct from other animal taxa. More significantly, the visibility of taxonomic categories in archaeological traces left by various activities in Phase 3 also denotes that animal categories were not only related to purely biological knowledge, but also had social connotations in a cultural context.

4.4.4 Spatial association between pig and longhouse

Pigs in general were spatially associated with longhouses. Pig-rich deposits were always located fairly close to longhouses, especially in Phase 3, where they all lay (pit H54, H35, H36, F2 animal pit and F8 animal pit) in the proximity of longhouses (F2 & F8) in the north of the excavation area (**Figures 4.9-4.11**). In Phase 2, pig-rich deposits composed part of the urn burial W1. Though there was a longhouse (F6) in Phase 1, no pig-rich deposits were recovered.

The association between pig and house bridges two key concepts: 'domestication' and 'household'. In fact, it would be inappropriate to isolate one from the other, as the term 'domestication' itself implies 'taking animals into the home', as demonstrated by its Latin origin. Domestication has been much discussed in the context of Chinese prehistory (*e.g.* Yuan & Flad 2002; Flad *et al.* 2007; Yuan & Dong 2018). Much ink has been spilled on the human-pig relationship in Neolithic China (*e.g.* Larson *et al.* 2005; Pechenkina *et al.* 2005; Larson *et al.* 2010, Cucchi *et al.* 2011; Luo 2012). Liu & Jones (2014) review isotopic data on humans and animals of north China from the early Neolithic to Bronze Age and recognise the correspondence of dietary signal between human and pigs starting with the middle Neolithic, indicating commensalism. Isotopic studies at Wadian demonstrates a similar picture whereby both pigs and humans shared access to millet crops (Chen *et al.* 2015). Liu and Jones (2014) further postulate that rigid swine control indicated by fodder provision could have corresponded to the privatisation of household resources, which can be used as evidence to illuminate the growing importance of 'household' in economic and social life in the middle Neolithic north China (*ibid.*).

A 'household' is defined as a group of people who reside in one house structure and cooperate to tackle a variety of tasks (Wilk & Rathje 1982; Blanton 1994; Netting *et al.* 1984). The household, being a basic albeit critical social unit, appeared to gain increasing autonomy over time in Neolithic China (Shelach 2006). This trend, illustrated by Li (2010), was also manifested in the spatial layout of some settlements which tended to be more 'close' and consisted of 'close-in' dining and storage area from the middle Neolithic. It appears that Wadian also fits into this picture. Each house seemed to have been walled and to have formed a close unit separable from any other. Pits within houses could have acted as storage facilities owned by each respective household. Moreover, burials were placed within or next to houses rather than in a communal cemetery. The household as an independent unit was emphasised by privatisation of both space and resources. It was now that the domestic animal as a private resource was connected to household in terms of labour organisation, food production and prestige display.

There is one exception where the pig-rich deposition was associated with urn burial (W1) instead of longhouse. Hodder (*cf.* 1984, 1990b) proposes that burials can be perceived as a metaphor for the house. This idea provides one possible solution to explain this exception.

At Wadian, therefore, the category of pig was connected to 'domestication' and 'household' through the bridge of materialised 'house'. By introducing pigs into domestic or household life, a changing living arena for pigs simultaneously implies a changing in their relationship to human and 'household'. In life, people lived with pigs under one roof. In death, pigs were deposited within or next to houses. The story continued across the domains of the living and the dead. The resemblance in deposit location of pigs and humans mirrored the view of Russell (2007) on animal domestication as kinship extended beyond human beings to other animal species.

Pig-household proximity at Wadian also directs us towards an examination of the character 家 *jia* (home), appearing in the seventeenth to eleventh centuries BC, which seems to depict a pig-like animal under a roof (Wang 2006; Liu & Jones 2014). The etymological interpretation of the roof was tracked to the radical 宀 mián (roof, Kangxi radical 40) which forms a range of characters whose meanings concentrate around 'home' 'household', 'house', 'family' and so on (*ibid.*). Though the animal underneath the roof sometimes appears to be dog instead of pig, the proximity between the animal and the roof shows a relationship of sharing (the house with human) and control (under the household), both of which can be encapsulated under a broad rubric pertaining to domestication and household.

4.4.5 Subdivision within 'pig' category

As it has been emphasised throughout the book, classification is deeply embedded in historic and social context. Under different contexts, people employ different conceptual organisation dealing with the ordering of things. Consequently an animal can simultaneously belong to multiple categories depending on conceptual organisation employed in the particular context. The section above demonstrates that the category of pig roughly resonates with the classification of pig in scientific biology, being distinct from other animal taxa such as deer. The pig category at Wadian, however, did not terminate at this level. Sub-division of pigs is also evident at the site.

Age binary classification

A further investigation into the characteristics of pig-rich deposits points out a subdivision of pigs according to their age. Two animal pits (F2 animal pit & F8 animal pit) within the longhouses were filled with juvenile pigs, suggesting a subdivision of pigs according to age group in foundation deposits. Foundation deposits are distinguished from other types of deposits in several ways. First and foremost, isolated juvenile pig remains were only present in foundation contexts. This trait has not been discovered in other types of depositions. Secondly, while only five out of 14 deposits have pig remains over 90 per cent, both F2 and F8 animal pits contained more than 90 per cent pig remains. In fact, every identified specimen in either context is from the pig.

Age appears to be a critical classificatory filter. Age categories were constructed upon a binary structure: the adult versus the young.

What purpose did this criterion serve? Animals placed beneath building foundations are usually interpreted as sacrifices used in ceremonies regarding the construction of buildings, while animal deposits adjacent to houses could be later sacrifices praying for the continuous prosperity of the house and the household in Neolithic China (Shi 1976; Yuan & Flad 2005). Provided this hypothesis is accepted at Wadian, Phase 3 pits might fit a pattern where F2 and F8 animal pits with juvenile pig remains were constructed to celebrate the establishment of the house and other pig-rich deposits, most probably, in pit H35 and H36, which would have been remains of later ceremonial events (*e.g.* feast) during the occupation of the house. Age-based pig differentiation supports the hypothesis of two types of ritual offering: one for construction of houses, the other for their maintenance.

The significance of age categories in ritual practices in prehistoric China during prehistory has been little explored. During the Zhou dynasty, preference for young animals as sacrifices was mentioned in textual records (Liu 2014). We have no reason to imagine any direct link between the Zhou sacrificial system and the Wadian pig category, considering the huge time gap. But the possibility that ritual performance regarding animal categories recorded in literature of later dynastic periods had prehistoric roots cannot be dismissed. This particular issue will be discussed further in chapter seven.

Despite the symbolic significance of the young animal category, use of juvenile pigs is of practical advantage from a more pragmatic point of view. When the rate of investment-and-return was taken into account, offering juvenile pigs instead of adults turned out to be an alternative method of cost minimalisation - less fodder and time had been spent on the juvenile, which provided less meat but also required even further care to ensure future growth. Balancing investments and returns, the economic cost of raising a young pig seems fairly high. Following this line of thought, a possible explanation of those pit deposits that contain adult pig remains such as pit H54, H36 and H35 is that those adult pigs were eaten and their bones were therefore food refuse. This hypothesis is in keeping with a theory of two types of ritual performance, one where ceremonies organised during the occupation and involved consumption of pigs perhaps in the form of feasting. Animals used in these ceremonies were not offerings or sacrifices to certain forms of higher power, and their consumption may have been prohibited under such situations. On the contrary, if consumption of animal meat was permitted, residents would have benefited from using adult animals. The distinction made in using animals of different age categories for different types of ritual ceremonies could therefore also be supported by economic considerations.

Domesticated versus wild

The Wadian category of pig does not seem to resonate with the dichotomy of 'wild' and 'domestic' inferred from the patterning of pig depositions. Both morphologically domestic pigs and wild boar are being discarded indifferently within one single context. No depositional contrast was discerned between 'wild' and 'domestic' status in the pigs from Wadian. Paradoxically, the asymmetric emphasis upon the 'domestic' side is clearly demonstrated, pigs (including wild boar) being associated with the house and household. On the one hand, the borderline between 'wild' and 'domestic' is not manifested in depositions, while on the other hand, 'pig' as a gross category regardless of age and husbandry status was always spatially connected to houses, underlining its symbolic attachment to the 'domestic' sphere.

The domestic pig, without doubt, might have been a member of the household, considering its proximity to human in terms of diet and living space. Explaining the presence of wild boar becomes tricky. Hodder (1990b) envisaged a conceptual process of 'domestication' where people's perception of animal status rather than animals' intrinsic properties is valued. Given a similar situation at Wadian, using the category wild boar is inappropriate in the biological sense. The wild boar could have been perceived as 'domesticated' through being brought into the household sphere of the community, and a wild category therefore need not have existed. Alternatively, even if a category of 'wild boar' existed, it may have been covert, at least in the utilisation of animal resources and deposition of animal remains.

Cross-classification of pig

Thus far I have repeatedly used 'pig' when referring to both domestic pigs and wild boar. This analytical unit at the Wadian site represents at least three osteologically identifiable groups: the group of adult wild boar, the group of fully-grown morphologically domesticated pigs, and the group of juvenile pigs lacking morphological information to identify wild from domesticated. 'Pig' in this context was clearly no terminal category but one that could be further subdivided. It remains debatable even whether these three groups can be grouped under a single hierarchical taxonomy: as noted previously, it remains possible that there existed independent categories from multiple classificatory systems operating in parallel.

The structure of a classificatory scheme is usually studied under the light of a nomenclature of assumed formalists in pre-industrial communities. But such nomenclature is inaccessible. If no available evidence supports any connection within one classificatory scheme, we cannot dismiss the idea that possibilities for cross-classification may be upheld at Wadian. Modern analogues provide evidence of cross-classification. Ross and Murphy (1999) employ taxonomic and script categories when approaching food categorisation, organising taxonomic categories the basis of food type (*e.g.* meat, vegetable) and script categories on the context of food consumption (*ibid.*). In line with these definitions, the taxonomic category at Wadian refers to the inclusion of animals based on similarities displayed in intrinsic characteristics (*e.g.* pig, deer, human in the broadest biological sense). Likewise, the situations where these animals were utilised underpinned script categories. Ross and Murphy (1999) postulate that script categories function as generating, rather than classifying particular items and therefore serve well to achieve a particular goal. I would envisage that pits containing juvenile pigs underneath house foundations resulted from sorting animals through the path of script categorisation in that a particular small group of animals (rigidly restricted) was generated as a consequence to accomplish sacrificial ceremonies.

The subdivision within pigs indicates at least two contexts through which different categorical systems operated. One dealt with rubbish. Separation of categories was not the top priority under this scheme because animals that ended up in this system first and foremost had lost their original value. That is to say, the natural characteristics, economic utilisation and social function attached to animals no longer mattered once they had been used and lost their intrinsic value. Consequently, animals were dumped altogether in refusal deposits regardless of biological traits, economic values, etc..

Another categorical groups was associated with ritual context. Age seemed to have been a critical factor in selecting sacrificial animals, as evidenced by the presence of foundation deposits containing juvenile pig remains. Wangchenggang site is the only other site in central China where complete skeletons of juvenile pigs in pits underneath house foundations (SAMPU & HPIACR 2007) have so far been found. Wangchenggang resembles Wadian in terms of both date and location. A detailed analysis of Wangchenggang is presented in the next chapter. Beyond this region, a handful of discoveries of juvenile pig remains associated with burial contexts have occasionally been reported (Luo 2012: 276-8).

Juvenile pigs were sometimes recovered in mixed deposits that contained pigs of different ages along with other animals. In other words, in certain circumstances, pigs were either categorised as animals of ritual importance or food waste. In practice, different classificatory systems might not have been mutually exclusive. Categories from different classificatory systems could have worked together simultaneously. Hence it is very likely cross-classification was widely practiced in the complex domain of animal exploitation at Wadian.

Animal individual versus animal product

Our final point about pig categories concerns the differentiation of items for classification. Linnaean taxonomy is concerned more with animals as biological entities. At Wadian, in addition to animals as individuals, animal products might have also been involved in categorisation. When animals are transformed from living individuals alive into animal products such as meat, it behooves us to re-evaluate the animal category.

A division of pigs on the basis of age (juvenile versus adult) rather than behaviour (domesticated versus wild) has been sketched out. This argument dwells on the presumption that categories of animal products (*i.e.* meat) corresponds to categories of animal individuals. For example, meat is categorised into pork, beef, venison corresponding to pig, cattle and deer respectively. Taphonomic study reveals that juvenile pigs in F2 and F8 animal pits might have been complete skeletons when interred. There were, on the other hand, traces of rodent gnawing (*e.g.* sample from H54) and weathering on pig bones deposited in other pits. As illustrated above, bones with gnawing marks were likely remains of consumption. Following this evidence, we could speculate that these remains were refuse from meat products. In the English vernacular, the animals in this scenario would betterbe rephrased as 'piglets' under house foundations as against 'pork' in refusal pits. The first problem here stems from the disjunction between animal parts as products and individual animals as organisms. Were animal products classified corresponding to the categories of animal organisms as biological entities? Or was there a separate system for animal products classification independent from classification of animal organisms? Dentan (1970) provides a simple solution, adopting 'two terminological sets, each of which contains terms that are homonymous with terms in the other' to interpret Semai classification of food and animals. Lacking further information, the question in Wadian remains open to exploration.

4.5 Summary of animal categories at Wadian

Taxonomic categories may have been established during Phase 1 and Phase 2 and drastically reinforced in Phase 3. They were beyond the domain of biological knowledge and widely employed in human life. Their embeddedness in daily life at Wadian corresponded to residential maintenance of social norms reflected in consistent site layout and the house construction. The category of pig was associated with houses, indicating attachment to household sphere. Nonetheless, the 'domestic' versus 'wild' contrast is not evident in Wadian depositional patterns: morphologically domesticated pigs and wild boar were found mixed in deposits. Age categories of pigs were recognised as probably serving ritual purposes. This categorical system emphasised the importance of situations where animals were exploited. As animal utility expired, they were dumped into the rubbish categories, regardless of biological and social characteristics. The distinction between the ritual and rubbish classificatory schemes fits the contrast between the two manners of deposition.

In short, cross-classification of animals might have been employed at the Wadian site. Different animal categories were facilitated with different economic and symbolic meaning, depending on different social contexts.

5

Animal Classification at Wangchenggang

Having examined the Wadian site in the previous chapter, the Wangchenggang site, dating from the Longshan through the Shang dynasty and situated about 50km northwest of Wadian, is explored in this chapter. This chapter begins by introducing the site before moving on to a critical review of previous faunal studies. A section of research results derived from both quantitative and descriptive data then follows, with a succeeding discussion section. A brief summary of categories at Wangchenggang comprises the conclusion of animal exploitation and perception.

5.1 Wangchenggang: site introduction

The Wangchenggang site, in Gaocheng Town, Dengfeng City, Henan Province, is a significant late Neolithic and early Bronze Age fortified settlement along the Ying River valley (**Figure 5.1**). Unless otherwise mentioned, the following introduction is summarised from two published excavation monographs (HPIACR & MCH 1992; SAMPU & HPIACR 2007).

The Wangchenggang site is situated on a mound adjacent to the Wudu River and Ying River. The mound was raised 1-2m above the surrounding land. Due to fluvial influence, the eastern facet of the mound was eroded, forming a sheer cliff.

Since its first discovery in 1954, several excavation sessions have been conducted at Wangchenggang. The first session in 1954 covered only a very limited area and most of the assemblage collected during that campaign was revealed to be Longshan remains. Excavations continued from 1976-1985. The most significant findis over this period constituted the discovery of settlement-enclosing moats and walls dating to the Longshan period. During 2002 to 2005, further excavations, as part of the nationwide 'Origins of Chinese Civilisation' project, were oriented towards exploring the role of the Wangchenggang site in its regional context. An additional circle of walls were unearthed outside of the walls previously discovered in the 1970s and 1980s. Apart from rich archaeological assemblages dated from the middle Neolithic Peiligang culture through early Shang Erligang period, the discoveries of more than one set of walls surrounding the settlement attracted the attention of archaeologists.

5.1.1 Chronology and phasing

Wangchenggang was occupied over a long period. The earliest archaeological can be traced back to middle Neolithic Peiligang, with discoveries of two pits and one tomb along with a small number of ceramics and lithics distributed in the west corner of the small enclosure.

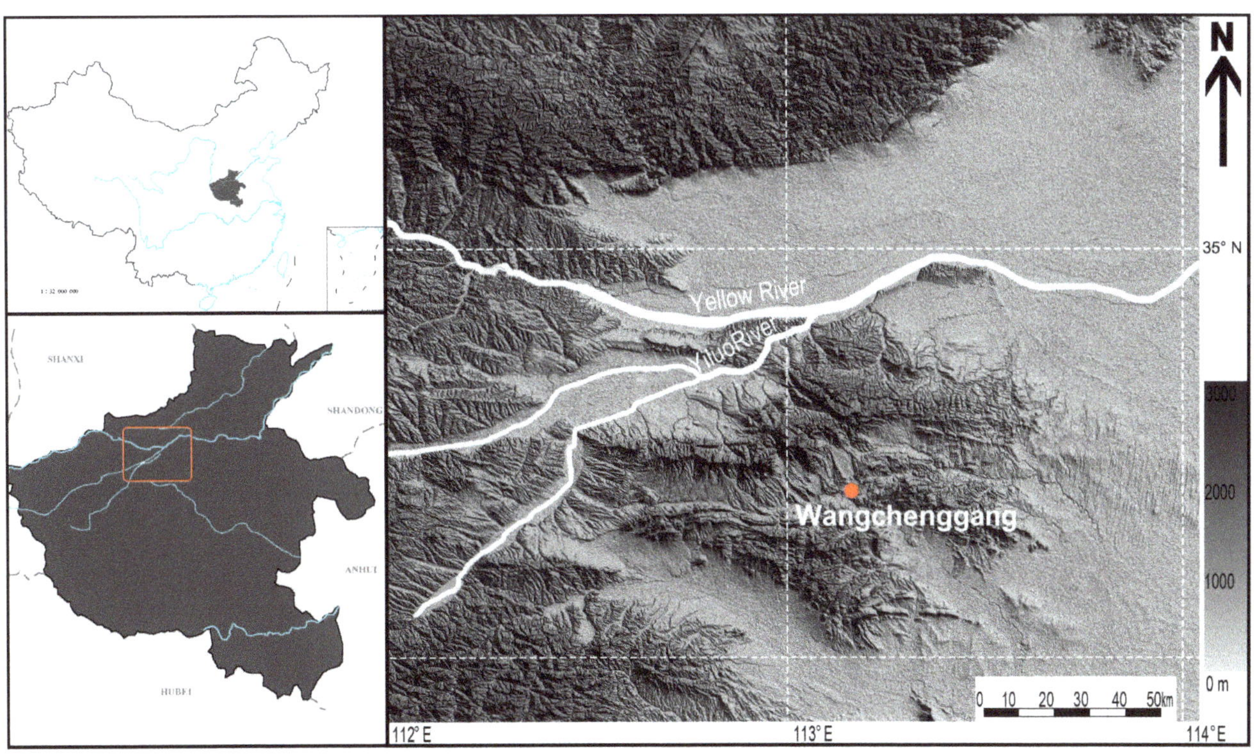

Figure 5.1 Map showing the location of the Wangchenggang site

Most features and assemblages date from the Longshan, Erlitou and Erligang (early Shang) periods. During the Longshan period, two walled settlements were recovered, containing a variety of pits and tamped earth, together with a substantial volume of archaeological assemblages. The Erlitou period witnessed the abandonment of the wall. The settlement area, however, remained in use, occupied by pits and tombs. When the Erligang period began, a large number of pits were constructed and were found containing rich ceramics, lithics, jade, bronze artifacts and so on. Over ten tombs were found dating to this period. The Yinxu period did not yield many remains and its distribution also retreated to the area of the former smaller enclosure. These cultural layers were covered by later dynastic depositions beyond the scope of this research period. A timeframe spanning the late Longshan period to the early Shang dynasty is the main concern of this study.

Multiple terminologies were employed to describe the chronology of different excavation campaigns. **Table 5.1** integrates dating information from reports of different excavations and attempts to piece them into a comprehensive chronological framework. The general sequential succession of the site is clear. The smaller walled settlement was constructed and used consistently throughout the late Longshan. A relatively large number of pits were recovered within the enclosure. The finding of tamped earth within the enclosure was also characteristic of this period. As the excavation was concentrated within the enclosure, little is known about the peripheral region. Abandoned at the beginning of the second millennium BC, a larger set of walls was erected at Wangchenggang, expanding the settlement to the southwest. Regarding the massive area enclosed by the wall, the excavation area was distributed around a rammed-earth wall, where a large number of pits have been excavated. Though the walls fell out of use around the terminal Longshan, the area previously enclosed by the wall remained in occupation. House structures along with pits and tombs dating to the Erligang period and onwards were revered inside

Wangchenggang seems to have had a complicated occupational history. The chronological sequence to date been sketched in very broad brushstrokes. A detailed discussion on the successive changes in settlement layout follows.

5.1.2 Spatial layout and feature types

The Wangchenggang site is characterised by its fortifications. The small walled enclosure was comprised of two parts separated by a north-south aligned wall. The large walled enclosure was constructed along the south and east of the small enclosure, the north wall cutting into the northwest corner of the small enclosure (**Figure 5.2**). Both fortified settlements date to the Longshan period.

The small walled enclosure was constructed and remained in use during Phase 2 of the Wangchenggang Longshan period (WCG Longshang Phase 2). The enclosure was comprised of two sections. Though the west section was preserved in better condition, the east was seriously deteriorated as a result of flooding from the adjacent Wudu River. The remaining south and west walls of the west section was 82.4m and 9m long respectively. The wall separating these two sections remained 65m long.

The major part of excavations in the 1980s covered the area in the west section of the small enclosure, where a large number of pits along with rammed earth remains

Table 5.1 Brief chronology of Wangchenggang and major feature types in each phase

<table>
<tr><th></th><th>1977 excavation</th><th>2002-05 excavation</th><th>C14 dates (BC)*</th><th>Features</th></tr>
<tr><td rowspan="5">Longshan</td><td>Phase 1</td><td rowspan="2">Early stage episode 1</td><td rowspan="2">2200~2082
|
2110~2045</td><td></td></tr>
<tr><td>Phase 2</td><td>Small enclosure wall, pit, rammed earth accumulation</td></tr>
<tr><td>Phase 3</td><td>Late stage episode 2</td><td rowspan="3">2130~2075
|
1855~1835</td><td>Large enclosure wall, pit</td></tr>
<tr><td>Phase 4</td><td rowspan="2">Late stage episode 3</td><td>Abandoned large enclosure wall, pit</td></tr>
<tr><td>Phase 5</td><td>Pit</td></tr>
<tr><td rowspan="4">Erlitou</td><td></td><td>Phase 1</td><td></td><td>Pit</td></tr>
<tr><td></td><td>Phase 2</td><td></td><td>Pit</td></tr>
<tr><td></td><td>Phase 3</td><td></td><td>Moat, house, tomb, pit</td></tr>
<tr><td></td><td>Phase 4</td><td></td><td>Pit</td></tr>
<tr><td rowspan="2">Erligang</td><td></td><td>Lower layer</td><td></td><td>Pit, tomb</td></tr>
<tr><td></td><td>Upper layer</td><td></td><td>Moat, pit, tomb</td></tr>
<tr><td>Yingxu</td><td></td><td></td><td></td><td>Pit</td></tr>
<tr><td colspan="5">* the C14 dating of each phase is derived from the earliest and latest dates among all samples belonging to the phase.</td></tr>
</table>

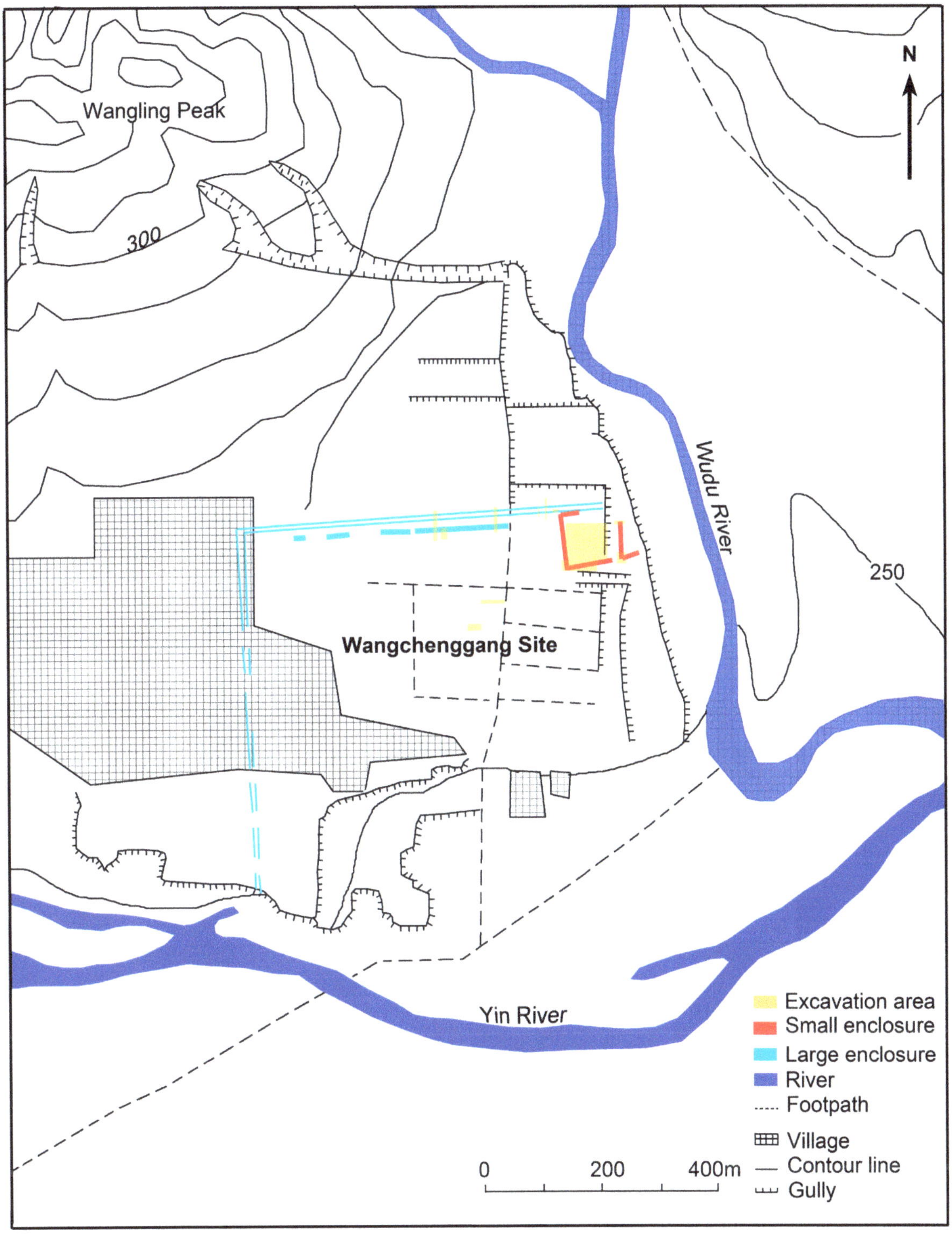

Figure 5.2 Schematic map showing the location of the enclosures at Wangchenggang (modified from SAMPU & HPIACR 2007: figure 3 & 4)

dated to Longshan Phase 2 and Phase 3 were found (**Figure 5.3**). The original site report (HPIACR & MCH 1992) marks out some of those pits that were filled with layers of stamped earth and contained human skeletons, coding them as 'foundation pits'. 13 Longshan foundation pits were found, containing articulated human skeletons (HPIACR & CHM 1992). A brief summary of these foundation pits is presented in **Table 5.2**. These pits were scattered in the western half of the small enclosure where a number of rammed earth accumulations and pits filled with tamped earth were discovered (*ibid.*). They were not neatly aligned, and some of the pits were clustered closely together. The original report indicates that these features together comprised a number of large structures built upon rammed earth foundations. At least 10 such features could be recognised (*ibid.*). However, it is difficult to spot the proposed building complexes. The spatial layout seems rather too random to form any architectural grouping.

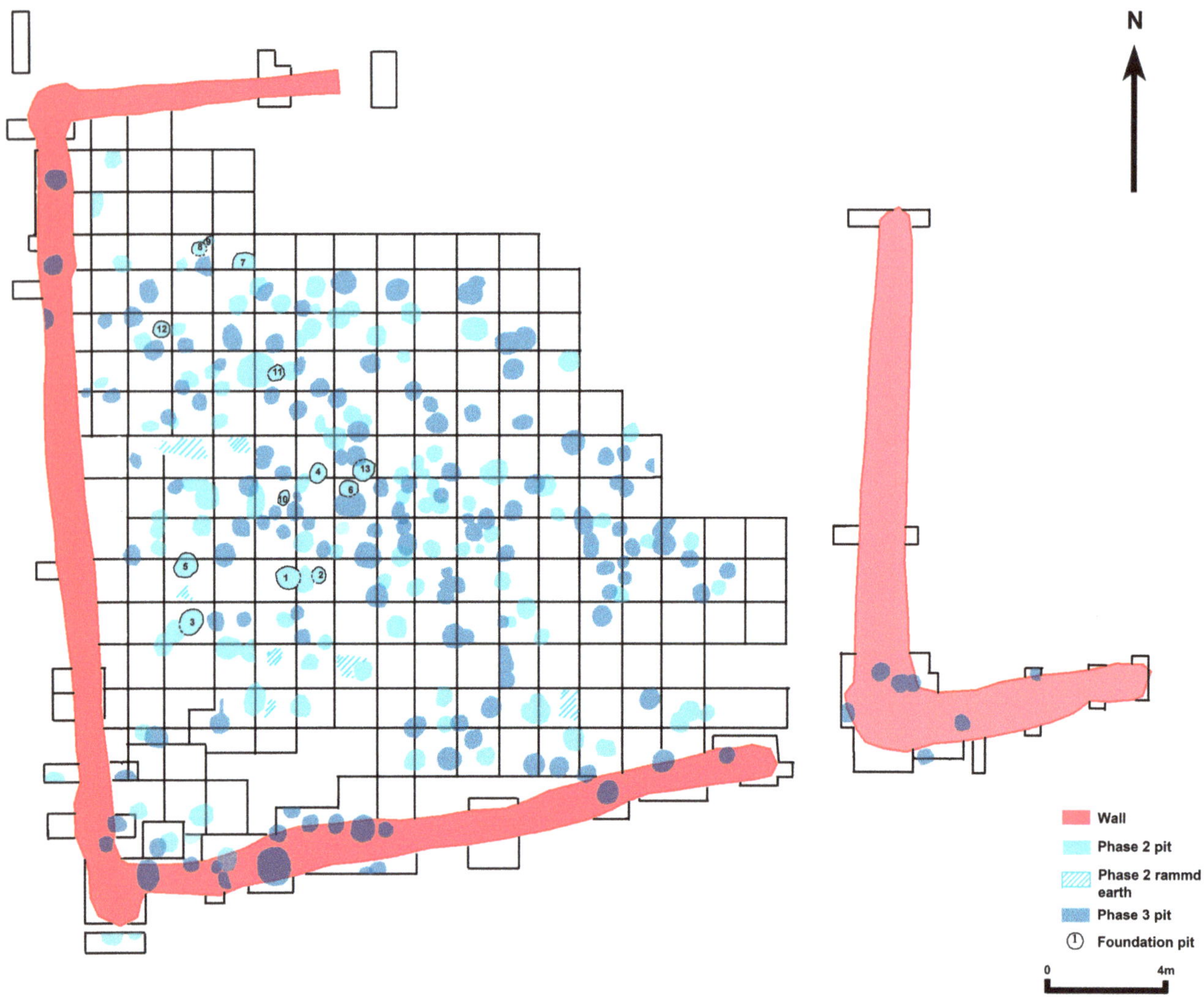

Figure 5.3 Excavation map of the Longshan small enclosure at Wangchenggang (modified from HPIACR & MCH 1992: figure 18)

Taking foundation pit 1 as an example, this was a several meter deep bag-shaped pit with a round opening (**Figure 5.4**). Twenty layers of tamped earth in total were identified in the pit. Seven human skeletons in total including adults, young adults and children were recovered from the lower layers. A child was interred in the third layer from the bottom and an adult male skeleton was found in the fourth layer from the bottom. One young adult female skeleton, along with an adult male skeleton, was deposited in the fifth layer. One adult female and two child skeletons were recovered in the sixth layer. Fragmented pottery was also found in the foundation pit, apparently mostly dining vessels such as *ding* (tripod), *guan* (jar), *jia* (tripod cup), *bo* (bowl), *zeng* (steamer), *dou* (high stem plate), *chenglüqi* (filter) and *hu* (pitcher). A small number of bones from pigs and deer together with stone tools were also discovered. The remaining twelve foundation pits, though varying in size, contained human skeletons and fragmented ceramics in layers of tamped earth.

The nature of these so called 'foundation pits' has been debated. The excavators associate them with construction rites, believing that a number of pits together with the rammed-earth fragments scattered among them might have constituted several large structures. This postulation, however, is challenged by Wang (1999) who points out that the spatial pattern of both pits and rammed-earth remnants was random and thus insufficient for a layout of large structures to be restored. Besides, due to the poor preservation, no clear evidence has thus far been unearthed supporting the idea that rammed-earth fragments and pits filled with pounded earth were components of larger structures. Therefore, we have reached no concrete conclusion regarding function and meaning of these foundation pits. In brief, it seems that the small enclosure was densely occupied during the Longshan, but its layout somewhat lacked any organised spatial pattern. The feature type tended to be uniform, being predominated by pits. Both features and remains decreased dramatically after the Longshan period.

Table 5.2 Summary of foundation pits in the small Longshan enclosure at Wangchenggang (from HPIACR & CHM 1992)

Number	Code	Opening Shape	Size (opening x bottom x depth)	Skeleton	Note
1	WT48H760	Round	2.52x2.94x2.66	7 (3 children, 2 male, 2 female)	In different stamped layers
2	WT48H120	Round	1.8x1.78x0.44	1 skull	
3	WT132H349	Round	2.3 unknown	1 male adult	Connected and cut into by rammed earth pits H347 & H348
4	WT254H419	Round	1.6 unknown	1 (probably adult)	
5	WT196H353	Round	2.25x2.6x1.51	2 adults	
6	WT157H416	Oval	2.04x2.2x0.7	1	
7	WT178H451	Round	1.0x unknown x 1.06	1 (probably adult)	
8	WT179H465	Round	1.84x1.73x0.2	Pelvis & lower long bones (ca. adult)	
9	WT179H466	Round	Bottom 0.2m destroyed	Pelvis frag	
10	WT195H476	Round	1.7 opening	1	
11	WT205H500	Oval	3.7x3.76x0.8	1 (without feet bones)	
12	WT240H607	Round	1.36x1.14x1.02	2 (2 adults; one most upper part; another lower part)	
13	WT233H571	Round	2.32x2.72x1.76q	5 skulls children; bone frags	

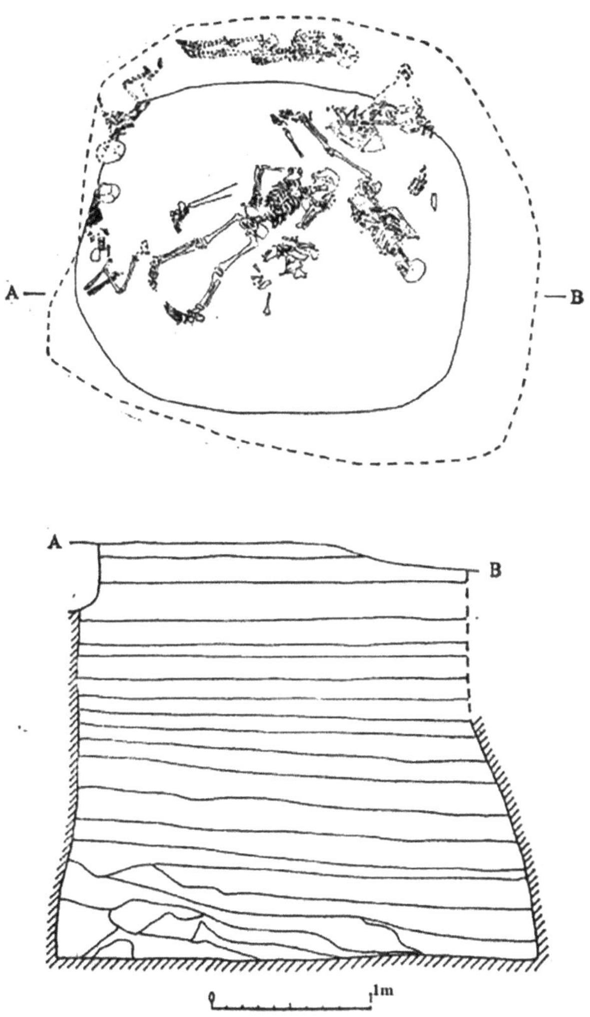

Figure 5.4 Plan and section drawings of foundation pit 1 at Wangchenggang (HPIACR & MCH 1992: figure 21)

The wall enclosing the larger area was latterly built during the Longshan Phase 3. The settlement expanded to the west and south, forming a larger enclosure of approximately 300,000m^2. The remaining northern wall was 350m long and 0.5-1.2m high, surrounded by a deep moat of 620m length, 15m width and 5m depth. The west moat remained 135m in length, 15m in width and 1.5-2m in depth. Other parts of the wall and moat were not preserved. The construction of the large enclosure began with the demolition of walls comprising the previous small enclosure. We see evidence of this from the intersection displaying the north wall of the large enclosure slicing into the northwest corner of the small enclosure wall. The area containing the former small enclosure was in the northeast quarter of this larger enclosure. The area remained in use during the Longshan Phase 3 occupation at the large enclosure , although only a handful of features and remains were recovered. It remains unclear whether changes in area function occurred alongside the disappearance of foundation pits in Phase 3.

With only a very limited area within the large enclosure excavated, sketching the entire picture of the spatial layout within the settlement is impossible at present. We receive the general impression of an overwhelming volume of Shang remains (Erligang & Yinxu periods) as compared with those from the Longshan period, even though the enclosure wall had been built and was functional during the Longshan period. The peripheral distribution of excavating trenches (mainly alongside the wall and in the north section of the site) may account for the imbalance in archaeological finds. Aside from this technical reason, examination of the spatial arrangement within the large enclosure awaits new evidence.

The excavation, conducted in the 2000s, was arranged in roughly two sections: one alongside the north wall of the large enclosure, the other within the centre of the north half of the large enclosure.

Wall-related excavation trenches contained features and remains dating to the Longshan, Shang dynasty and later Eastern Zhou dynasty. Excavation maps of each area are presented in **Figure 5.5, Figure 5.6** and **Figure 5.7**. Pits in general are the predominate feature type throughout all three phases. Feature types gradually diversified towards the historical period. A pair of Erligang burials were recovered adjacent to the north wall. These pits contained, variously, ceramic fragments, animal bones and human skeletons. A sheep diviniation bone dating to the Longshan period was also excavated from a pit adjacent to the wall.

Figure 5.8 provides a map of the excavation area within the large enclosure. The wall of was also constructed during the Longshan. By strong contrast, for some reason, very few Longshan features were discovered within this area. The relatively diverse features, including pit, house and tomb dating to the Shang dynasty (Erligang and Yinxu periods), were unearthed. Dating to the Erligang period, the pile of modified human skulls discovered around excavation squares W5T0344 and W5T0544, is worth mentioning, and was likely a workshop for manufacturing skull-cups.An Erlitou period deposition containing one adult pig and one juvenile pig within the house structure was also found in trench 85WT264.

(Paragraphs above summarised from HPIACR & MCH 1992 and SAMPU & HPIACR 2007)

5.1.3 Wangchenggang and its regional network

Fortified settlements increased dramatically during the late Longshan period. As described in chapter three, a variety of terms such as 'state' have been employed to describe the economic, political and social scenarios Longshan central China faced. Regardless of the most accurate term, the existence of settlements with defensive structures of varying degrees is evident. Wangchenggang, a settlement fortified by walls and ditches, was one such settlement.

Many scholars believe that the Wangchenggang walled enclosure might have been a pivotal place in the regional network in the late Longshan period. Fang (2002), having participated in multiple excavations across this region, recognises the hierarchical pattern in the upper Ying river valley during the Longshan period – an area where over 30 Longshan culture sites were distributed. The Wangchenggang site might have served as one of the centres, surrounded by lower ranking sites including Shiyanguan, Baijiacun, Yuancun and Yangcun. This hierarchical pattern came to be seen as multi-centred where the Wandian site was also recognised as a centre surrounded by smaller sites such as Yanzhai, Wuwan and Gushuihe *etc.* (*ibid.*). Fang (*ibid.*) listed several additional criteria for distinguishing central settlements including settlement size, defensive structure, presence of ceremonial pits, construction of rammed-earth architecture, production of bronze artifacts, utilisation of oracle bones and so forth. Not all these features, however, can be found at both Wangchenggang and Wadian. For example, the Wadian site was not fortified by walls or ditches.

As the Wangchenggang site ranked at the top of its regional settlement network, some Chinese archaeologists have associated the site with the legendary capital of Yang (*陽城 Yangcheng*) belonging to the Emperor Yu (*禹*), as recorded in ancient literature. This follows an historiographical tradition (*e.g.* An 1985; Ma 2008). This assumption runs the risk of treating historic records as reliable reference points to archaeological discoveries.

When the enclosure wall was entirely abandoned during the Shang, the settlement declined to the leel of a village like any regular residential settlement along the Ying River valley. By this time the regional centre had moved eastward to the great Shang settlement at Anyang.

While most studies focus on the Longshan settlement hierarchy, highlighting the pivotal place of the Wangchenggang site as the regional centre, we should also examine the chronological site ranking, especially considering the site's long occupation. Enclosing walls that constructed and used during the Longshan period, however, were soon abandoned in the Erlitou and Erligang periods. It is plausible there were possible changes in the nature and the ranking of the site in the regional system before and after the abandonment of the defensive structures.

5.2 Previous zooarchaeological studies

A preliminary analysis on the faunal remains from all phases (including dynastic periods) excavated during 2002-05 seasons was attached to that excavation report. According to the report (SAMPU & HPIACR 2007), a collection of 7428 pieces of faunal remains was excavated among which 1461 pieces of bone were only identified with very broad taxa such as large mammal, medium mammal, small mammal and mammal. The remaining bone pieces were counted as NISPs. Up to 3511 pieces, however, belong to *Cipangopaludina cathayensis,* a freshwater snail taxon. In terms of taxonomic abundance among mammals, pigs predominated the assemblage throughout the phases, though slightly declining in number during the Erlitou and Erligang periods before a resurgence around the Yinxu period. The trend in cattle was exactly the opposite, with the number of remains increasing from the Longshan to Yinxu. Dogs, deer and sheep were also present in fauna assemblage.

A revised NISP count will be calculated and presented in the results section, based on which further analysis will be performed according to the method proposed in this project.

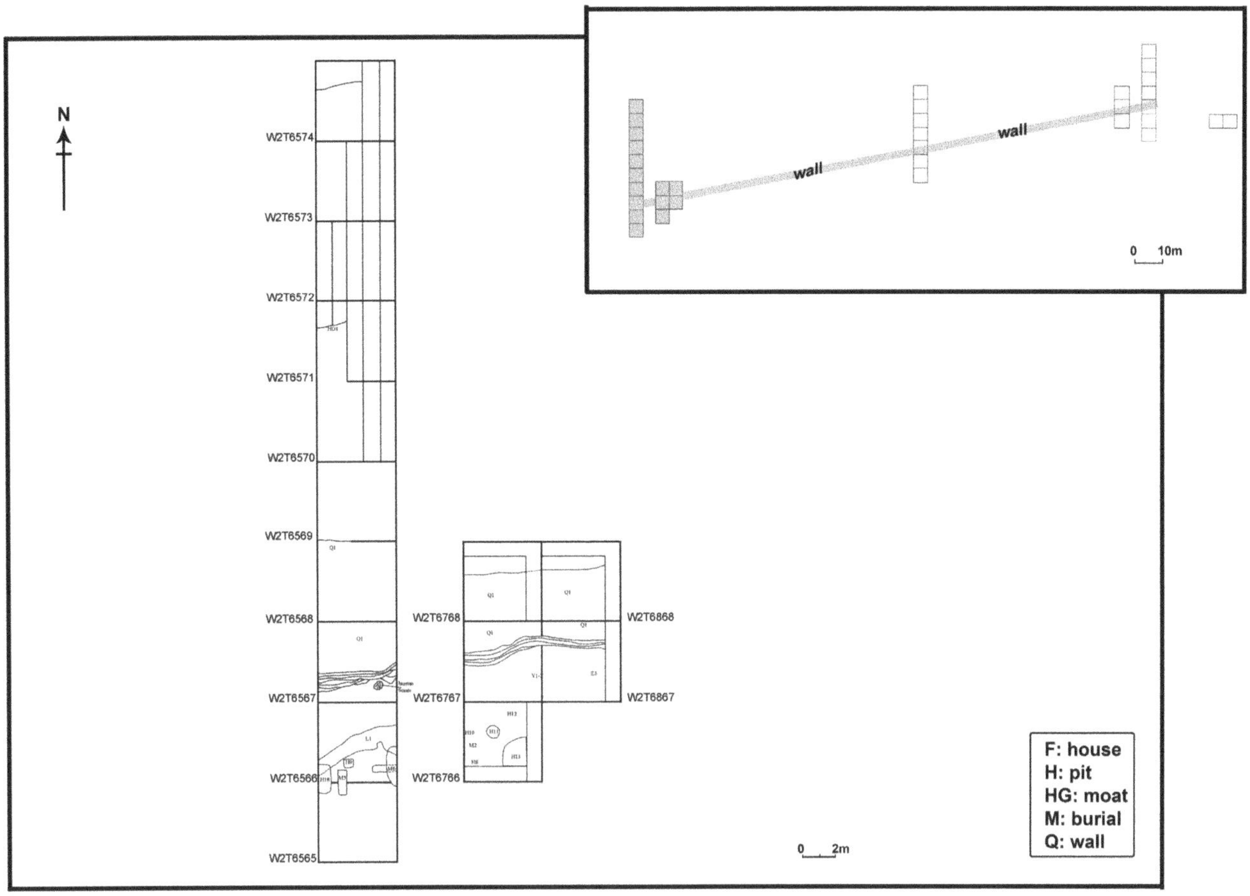

Figure 5.5 Excavation map of the trenches adjacent to the enclosure wall in the west at Wangchenggang (modified from SAMPU & HPIACR 2007: figure 6 & 8)

In addition to the preliminary assessment above, further steps taken to explore animal husbandry from a regional scope also include book sections dwelling on Wangchenggang fauna. Yuan and colleagues (2007) integrate faunal information from several late Neolithic and early Bronze Age sites in central China (including Wangchenggang) and propose that the animal exploitation strategy practiced at Wangchenggang broadly agreed with the regional institution inherited since the middle Neolithic, in which domestic animals, largely pigs, served as the major meat provision. At the same time, recently domesticated sheep and cattle began to appear in several sites (*ibid.*).

5.3 Results

3463 bone fragments were collected from the Neolithic and Bronze Age cultural layers. Among them, 439 specimens were mammalian. While mammal remains occupy a small portion, shell remains comprise the majority of the remaining assemblage. 185 mammal specimens were collected from feature depositions. The Erligang period yielded the most numerous faunual remains. This was followed by the Longshan period. Fairly small samples belong to the Erlitou and Yingxu periods. **Table 5.3** provides a summary of NISPs of each animal taxon collected from the site. Other animals include dog, hedgehog, rat and rabbit, to just name a few. The Erlitou and Yinxu assemblages are of very small sample size.

As the excavation trench was scattered over a wide area, the following analysis focuses on faunal assemblages deposited in feature contexts (*e.g.* pit, house and so on.) which more strongly indicate a relationship between deposition formation and settlement layout. The bar chart in **Figure 5.9** demonstrates the proportions of the four main animal taxa. Erlitou and Yinxu assemblages are excluded from comparison due to the extremely small sample size. Generally speaking, pigs constitute the major part to the faunal assemblage, followed by sheep and cattle. Deer remains are relatively fewer in number.

5.3.1 Animal taxa

Having sketched the general picture of the faunal assemblage, major animal taxa occurring at the site are discussed one by one in the following section.

Pig

As seen in **Table 5.3** and **Figure 5.9**, pig remains always contained the highest NISP counts throughout the four periods. Nonetheless their proportions dropped from the 72 per cent of the Longshan period to under 50 per cent by

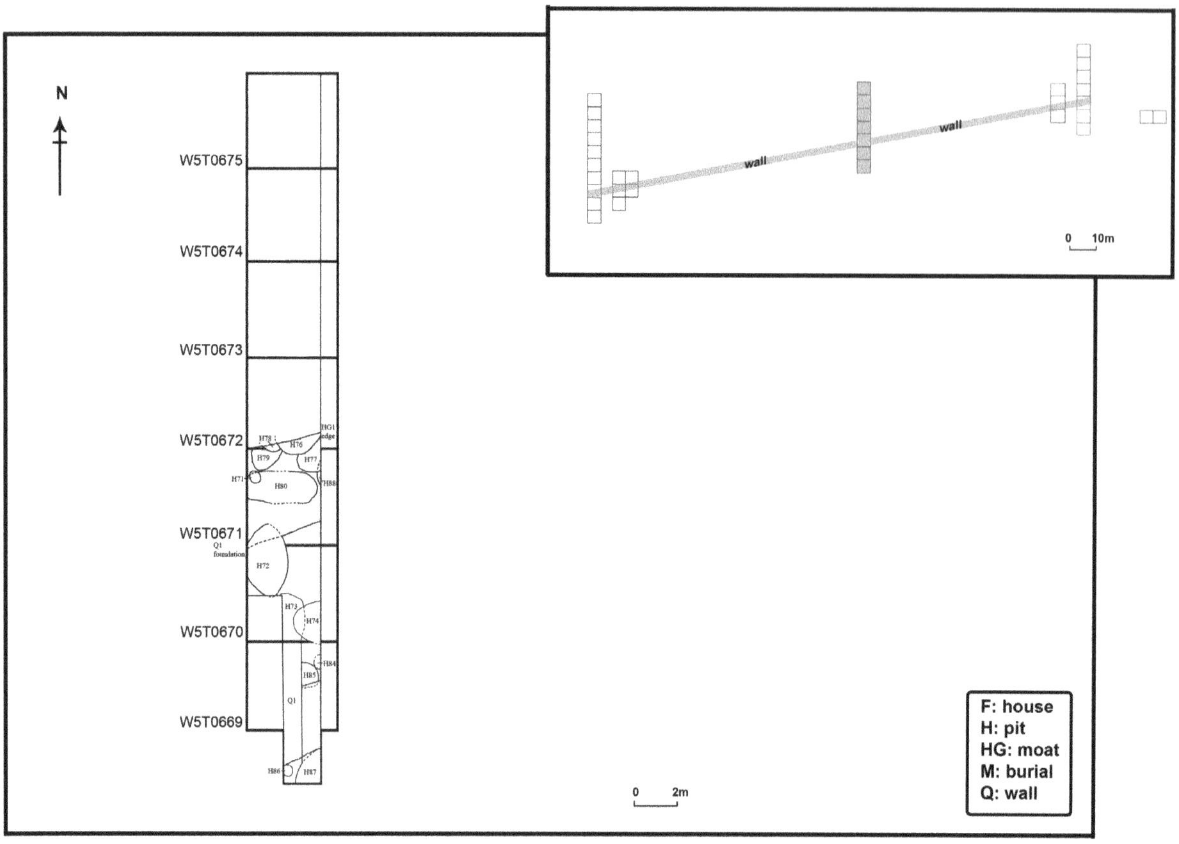

Figure 5.6 Excavation map of the trenches adjacent to the enclosure wall in the middle at Wangchenggang (modified from SAMPU &HPIACR 2007: figure 14)

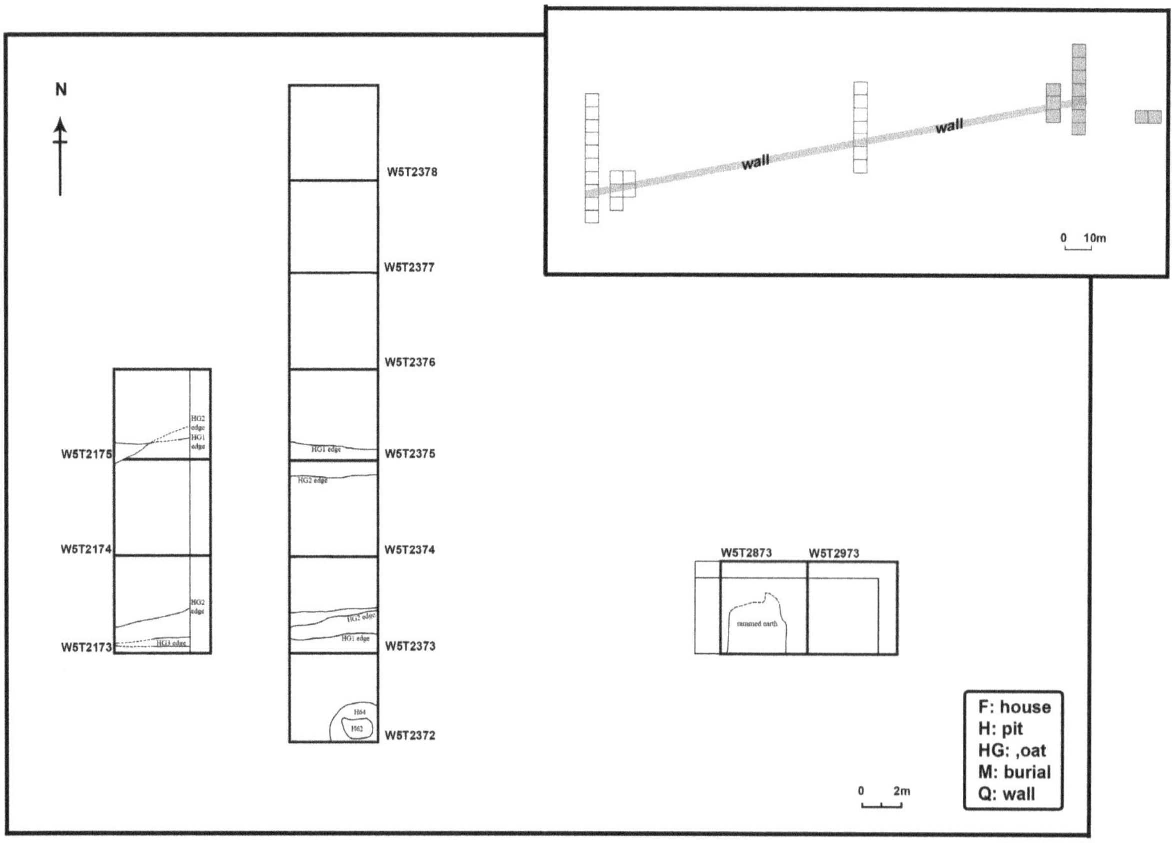

Figure 5.7 Excavation map of the trenches adjacent to the enclosure wall in the east at Wangchenggang (modified from SAMPU & HPIACR 2007: figure 16 & 18)

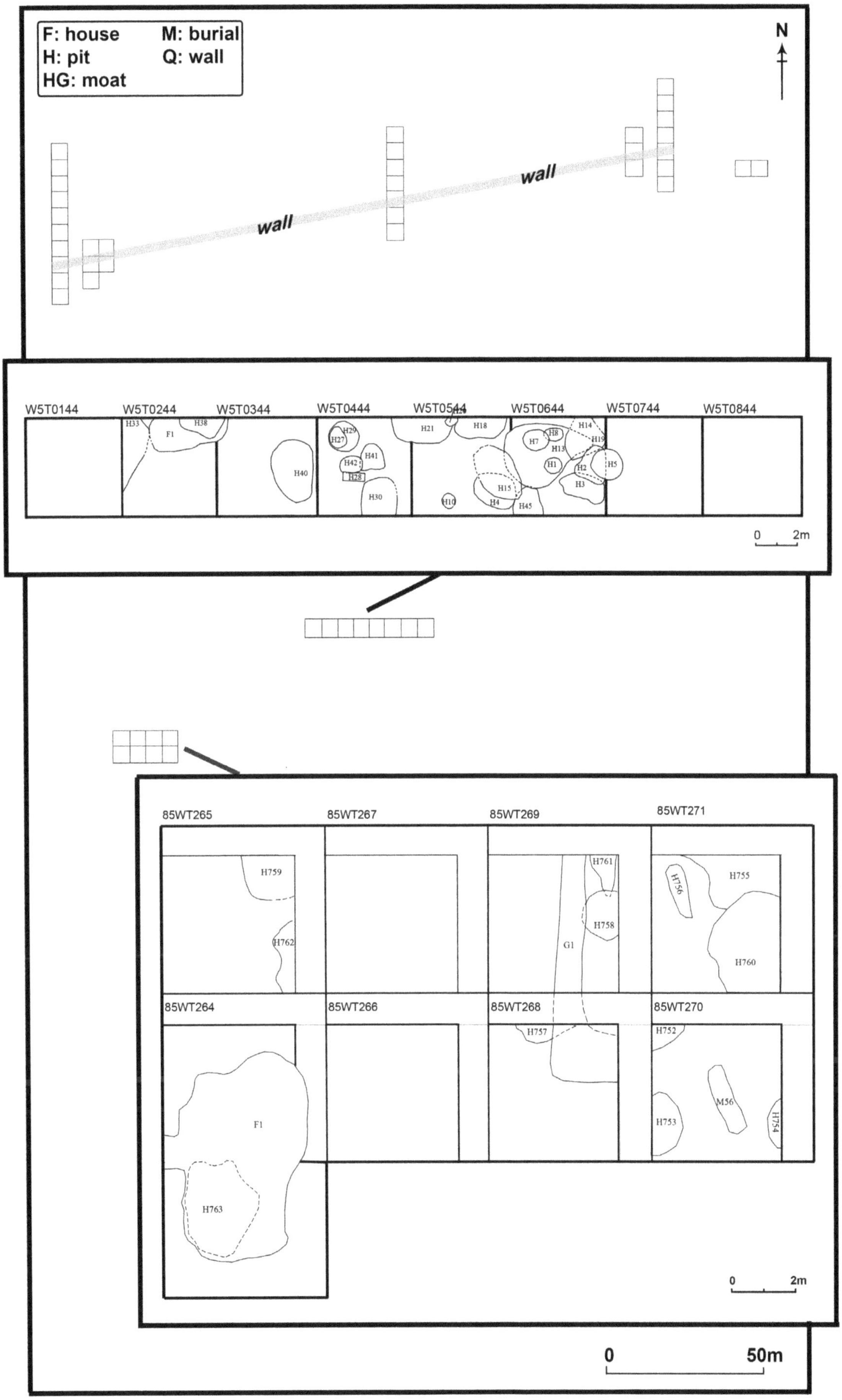

Figure 5.8 Excavation map of the trenches within the large enclosure at Wangchenggang (modified from SAMPU & HPIACR 2007: figure 10 & 12)

Table 5.3 Summary of taxa abundance at Wangchenggang

Period	Context	Pig	Deer	Sheep/goat	Cattle	Other animals	Total
Longshan	Feature	44	2	0	5	4	55
	Layer	23	0	7	2	6	38
Erlitou	Feature	2	0	2	1	0	5
	Layer	5	1	2	0	4	12
Erligang	Feature	68	6	8	24	12	118
	Layer	92	16	9	29	58	204
Yinxu	Feature	4	0	1	2	0	7
	Layer	0	0	0	0	0	0
Total		238	25	29	63	84	439

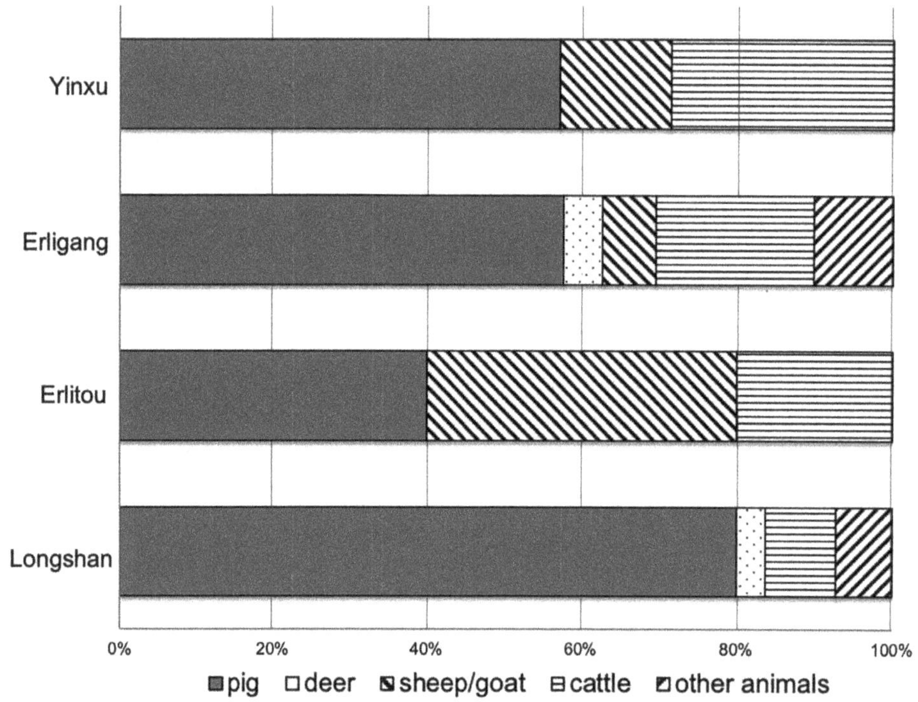

Figure 5.9 NISP percentage of major animal taxa at Wangchenggang

the Erlitou. A more precise temporal changing pattern is missing as a result of the highly limited recovery of Erlitou and Yinxu faunal remains.

Deer

Deer remains were the subject of occasional finds. All belong to sika deer (*Cervus nippon*). We found two Longshan specimens and 22 from the Erligang period, which have yielded remains of comparable sample size. This disregards the Erlitou and Yinxu periods, from which faunal remains were only occasionally collected. The slight increase does not make a major difference in proportion. Taking small sample size into consideration, it would be hard to reach any concrete conclusion regarding the nature of deer exploitation.

Sheep

Sheep remains were unearthed from layer contexts dating to the Longshan period. They only appear in feature depositions from the Erlitou onwards. The remarkable increase of sheep from this time largely consonant with the general dating of the arrival of domesticated sheep in China from West Asia, occurring in this region roughly during the second millennium BC (Tu *et al.* 1989; Chen 1994; Cai *et al.* 2007). In general, sheep bones were relatively low throughout all periods. Only four and one sheep specimens were collected from the Erlitou and Yinxu deposits. The proportion of sheep in each period is uneven. In those assemblages with comparable sample size, i.e. the Longshan and Erligang periods, the percentage of sheep NISP remains invariably under 10 per cent. For those small-sample-sized assemblages this may not precisely reflect the real distribution of animal taxonomic abundance. The more bones collected, the smaller the proportion of sheep. In the Erligang period where 322 pieces of bone were found, the proportion of sheep remains was 5.3 per cent, while in the Erlitou period where only four bones were found, the proportion of sheep had increased to 23.5 per cent.

Cattle

Cattle remains were present in all four periods. A relatively large volume was unearthed from Erligang deposits, along with the significant increase in the proportion of all faunal remains in that same period. Comparing the proportion in each period, proportional increase occurs from 7.5 per cent of the Longshan period up to 16.5 per cent in the Erligang period. The enlarged proportion of cattle seemingly indicates its increasing importance in human exploitation, as likewise reflected in patterns of sheep remains.

Other mammals

The remaining specimens include dogs and some small mammals such as rabbit and hedgehog. As only common names are used in the original report, exact species are not identified. A large proportion of specimens in this group belong to domestic dogs. The animal taxa present in this group were not diverse.

5.3.2 Presence of human bones

For the purposes of my analysis, human remains should be best recorded according to the same criteria for non-human animals. However, the original format and contents of skeletal information of human beings differs significantly from that of animals, making the same analytical process an impossibility. On one hand I must maintain a view of the human as a categorical unit indistinguishable from other animals, and on the other hand be honest about the limitation of my dataset. As information about human remains is mainly descriptive, in this section I integrate available information of human remains. Further quantitative analyses will be attempted if the format and resolution of the dataset allow.

The human bone data in this section is summarised according to HPIACR & CHM (1992). Most human remains were unearthed from the 1970-80s excavations and information was reported in the original site report. The skeletal information is descriptive. The following information is summarised from the HPIACR & CHM report, published in 1992:

Of the foundation pits from the 1970s and 1980s, pits dug and filled in advance of building or structure were the major feature type for deposits of human skeletons. Foundation pits deposited with complete human skeletons include H348, H419, H353, H416, H451, H476 and H500, situated within the west section of the Longshan period small enclosure. Except for the foundation pit 1, the remaining foundation pits were not fully excavated for the purpose of preservation. The locations of foundation pits are displayed in **Figure 5.3**.

Foundation Pit 1, coded as WT48H760, was a pit containing a round opening and round bottom. The pit was filled with stamped layers of earth within which seven individuals including three children, two adult males and two adult females were discovered. This is the largest human skeleton pit recovered so far. Foundation pit 2, coded as WT48H120, was located next to the foundation pit 1. Both were cut into by pit H106, which ran through the middle, thus connecting to each of them. Foundation pit 2 contained one skull. Although not clear in the report, the text suggests that it was probably a complete skull. The pit featured a round opening, with a 1.8m diameter. Foundation pit 3, coded as WT132H349, was located west of foundation pits 1 and 2. The opening was also round. A skeleton of an adult male was placed on top. The pit was cut into by pits H347 and H348, which were filled with rammed earth. Pit WT254H419 is foundation pit 4, with a round opening as well. One skeleton was found, whose age and gender could not be identified. Foundation pit 5, coded as WT196H353, was situated adjacent to foundation pit 3. Between them there was an accumulation of rammed earth remnants. The shape was similar to foundation pit 1. Two adult skeletons were buried in it. Foundation pit 6, unlike the majority of the foundation pits, had an ovoid opening. One skeleton of uncertain age and sex was found deposited in this pit. Foundation pit 7 contained one skeleton lying in an unusual deposition with the head placed higher than the feet. In foundation pit 8, only the human pelvis and lower long bones were collected. Foundation pit 9 had been seriously damaged. Although the round opening remained identifiable, the lower part had not survived. Fragments of a human pelvis were recovered. All were located in the north of the settlement. Foundation pit 8 and foundation pit 9 were cut into each other. Foundation pit 10 had a regular round opening, the diameter of which was approximately 1.7m. One complete skeleton was deposited in it. The gender and age of this skeleton are not clear. Foundation pit 11 contained one skeleton whose feet bones were missing. This pit was a large but shallow pit with an ovoid opening at the surface. Two skeletons were deposited in foundation pit 12, located in the west margin of the site. One skeleton survived with only the upper body part, whereas the lower body of the other was all that remained. Pit H375 cut into the foundation pit from the east. Foundation pit 13 was of a round opening with a pocket-like vertical section. It was situated next to foundation pit 6 and foundation pit 4. As many as five child skulls were excavated from the pit, along with a few scattered bone fragments.

(Paragraphs above summarised from HPIACR & CHM 1992)

During the excavation in the 70s and 80s human burials were also uncovered in the area between Wangchenggang and the Bafangcui site to the southwest (HPIACR & CHM 1992). As no map showing the distribution of features was attached to the original report, we can only tentatively infer that their locations were in the vicinity of trench coded with the initial '85W' during excavations in the 2000s (**Figure 5.2** & **5.8**).

Ten human tombs from the Erligang period were reported during the excavation, with detailed records provided for five of these. All 10 shaft burials were located in the open area to the northwest of the small enclosure. The following paragraph is summarised from HPIA & ADMCH 1992:

An adult male skeleton was excavated from tomb WT9M8. The skeleton with its head facing north was buried along with funeral goods comprising several ceramic vessels. In tomb WT19M11 the skeleton of an adult human was discovered, alongside a dog skeleton. This was placed west to the human body and a lithic tool placed next to the individual's head. Tomb WT16M14 contained a complete adult male skeleton with head oriented northwest. Funeral goods including pottery fragments, copper arrowheads, bone arrowheads, jade ornaments and shells were also found in the tomb.

In the recent excavation conducted during the 2000s, human remains were recovered from nine deposits (including pit, tomb, ditch and cultural layer). The following information is summarised from SAMPU & HPIACR (2007):

Pit H64 of the Longshan period, situated about five meters south to the northern ditch of the large enclosure, yielded a child skeleton. The skeleton was placed in the north of the pit, head to the north and face down. Underneath were accumulations of ceramic fragments from *ding* (tripod), *guan* (jar), *bei* (cup), *gang* (tank), *bo* (bowl), *weng* (urn) along with molding stones, lithic spades and stone ornaments. Arrowheads made of animal bones were also discovered. Tomb M56 is dated to the Longshan period. It was located format a relative distant from the ditches and walls, within the fortified enclosure. An incomplete adult skeleton lying on its back was recovered within the burial, its head facing southeast. The Erligang period Tomb M5 was situated several meters south of the northern wall Q1 in the west excavation part. The skeleton belongs to a three-year-old child, their head in the north part of the burial, and facing west. One scapula from an unidentified mammal was buried along with the child. Apart from skeletons properly placed in tombs, skull fragments of an adult female were recovered in one section of ditch HG1 in square W2T6572, and dated to the Longshan period. In another Longshan pit H74, human bones were discovered mixed with other animal remains. The excavation report makes brief mention of two incomplete skeletons excavated from the Longshan cultural layers, one belonging to infant and the other a female. The exact locations were not reported.

During the Erlitou period, a right human femur was discovered in ditch HG2 in square W5T2374. It was the only human remain excavated from this period.

In the Erligang period, an accumulation of human remains was found in the square W5T0544, probably near the centre of the settlement. There were 4 skulls in total clustered in the north part of the square: these belonged to three adult females aged from 20 to 40 years old and one unsexed adult. Chopping marks were tentatively identified on two skulls. The Erligang pit in excavation square W5T6566 contained several long bones. Tomb M6 in the same trench square also contained a deposit of an incomplete female skeleton in her later thirties. Pit H40, in square W5T0344, yielded skull fragments from an adult male. Pits H13 and H21 were filled with both human and animal bones. Their size was remarkable and they were located near the central part of the settlement.

(Paragraphs above summarised from SAMPU & HPIACR 2007)

5.3.3 Fragmentation

Fragmentation was evaluated through examination of identifiable ratios. Ideally, the fragmentation rates of deposits containing human remains ('sacrifice pits' in the original excavation report) would be included depositions very likely intentionally constructed to serve specific purposes. However, information on human remains is insufficient to perform a quantitative analysis. Analysis on fragmentation of human skeletons has to be put aside for the time being.

Seen from results (**Figure 5.10**) where contexts of all phases have been lumped together to demonstrate the general pattern of the fragmentation, the majority of the contexts (n=23) fall within the range from 20 to 60 per cent. The number of contexts within the the group of 90–100 per cent is surprisingly large, ranking the fourth amongst all percentile groups. It is also interesting to observe a gap, with no contexts falling within the 70-80 and 80–90 per cent groups. This pattern tentatively fits in a normal distribution with outliers scattering around the 90-100 percentile. A further examination into the sample size of each context shows a very small sample size for these contexts with higher identifiable ratios. There are several possibilities which may explain the distribution. Firstly, the risk of over-presentation due to small sample size cannot be dismissed. Secondly, the limited excavation compared with the large area which the settlement might have covered should also be counted. Thirdly, the bimodal distribution could indicate two distinct formation processes. With depositions from various periods are aggregated to consider their fragmentation altogether, such a distribution with outliers is also possibly a time-averaged illusion: one mode might have belonged to an earlier period while the other appeared in a later period. In this situation, there might have been two types of non-contemporary formation processes.

The fragmentation pattern for each phase was also examined. During the Longshan phase, the average range of identifiable ratio is roughly in 20-80 per cent, higher

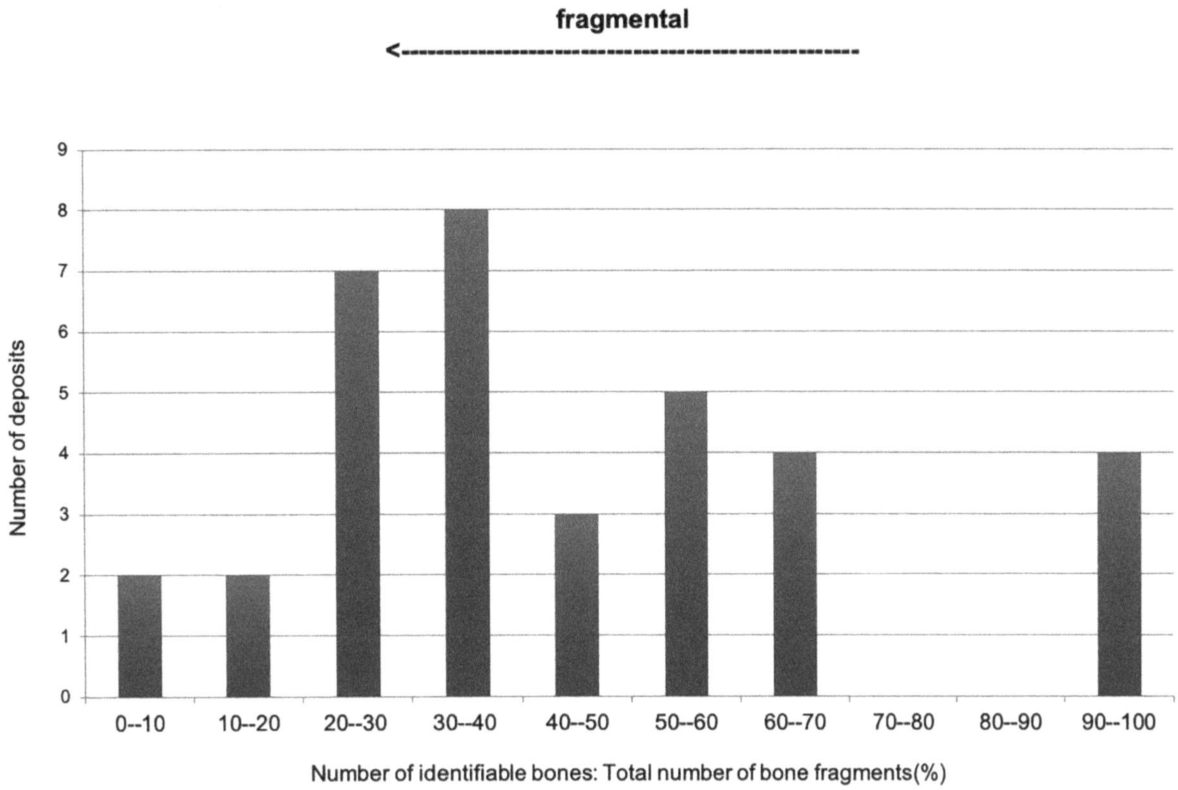

Figure 5.10 Fragmentation of faunal assemblages in depositions at Wangchenggang

than all other phases combined. This shift is largely due to the increase in the number of contexts in the 90-100 per cent group. Meanwhile the gap betweenn this group and the lower percentage groups is broadened, with no contexts falling within the 60-90 per cent group. During the Erligang period, the average identifiable range remains the same as that of the Longshan period, lying between the 20-60 per cent. The distribution, however, reaches the peak in the 20-30 per cent group and drops dramatically in the 40-60 per cent group, differing from the Longshan pattern. Only one deposit appears in the 90-100 per cent group, but none in the 70-80 and 80-90 per cent groups, extending the gap between them and the 90-100 per cent group. In both the Erlitou and Yinxu periods, the number of contexts present with an identifiable ratio is extremely limited, removing the possibility of comparison.

5.3.4 Body element distribution

The distributions of body element of pig, cattle, sheep/goat and deer from feature depositions were calculated and their results presented in **Table 5.5** and **Figure 5.11**. For comparison sake, we note that the Erlitou assemblage is extremely small, while the sample size of the other two is not large either. For human assemblages, information on body element is not provided. Hardly can any accurate facts be said about cattle and sheep/goat, due to their limited discoveries. Bones from head elements make up a large proportion of the diachronic pig assemblage. Relatively more pig limb elements were found than for other body parts. The chronological pattern remains unclear. For the deer assemblage, it seems during the Erligang period, the major contribution came from extremity elements, despite fairly low NISP counts. Only one or two deer specimens belonging to other body elements were excavated.

5.3.5 Associated bone group

Associated bone groups are bones belonging to a single deposition that might have come from one individual. Two non-human animal ABGs were found, both in the context of pit H76, and belonging to the Longshan period. One was constituted of five pig bones belonging to a juvenile individual and the other comprised two bones from a dog. We note that some of those human remains that were found in tombs and foundation pits essentially formed ABGs. Detailed information about human skeletons has been presented above.

5.3.6 Age profile

Lacking records such as tooth eruption and wearing, and epiphyseal fusing for aging, a systematic analysis on age of death is absent. The only descriptive information about animal age at death is accessible in the previously published faunal report (SAMPU & HPIACR 2007). According to the report (*ibid.*), one pig specimen from the Longshan period dated as a juvenile pig around one year old. Thirteen Erligang pig specimens were examined, 10 from pigs under one year old. The remaining three specimens were from pigs around half to one year old,

Table 5.4 Identifiable ratios of assemblages in each context at Wangchenggang (F: house; H: pit, HG: ditch)

Context	Identifiable ratio (%)	Number of identifiable specimens	Context	Identifiable ratio (%)	Number of identifiable specimens
85F1	57.1	7	H27	7.7	13
City wall	100.0	1	H29	22.7	44
F1	50.0	4	H30	8.7	23
H1	40.0	5	H41	33.3	9
H3	50.0	2	H45	25.0	4
H4	66.7	3	H47	45.5	11
H5	25.0	48	H49	100.0	3
H7	60.0	15	H62	20.0	5
H8	33.3	9	H72	30.4	23
H9	100.0	3	H73	57.1	7
H10	28.6	7	H74	25.0	4
H11	31.7	41	H76	39.7	73
H13	34.9	109	H77	50.0	2
H15	66.7	3	H79	100.0	2
H18	27.8	18	H80	42.9	7
H19	66.7	6	H85	33.3	9
H21	25.0	4	HG3	33.3	3
H26	15.4	26			

belonging to the calfing group (**Table 5.6**). The figure is strikingly skewed towards young individuals.

An age profile of humans has been attempted, despite the wide timeframe. These remains are broadly separated into two age groups: the young and the adult. As shown in **Table 5.7** and **Figure 5.12**, results was gauged for each period regardless of the difference in feature type, on an individual basis. The total number of adult individuals twice the number of young individuals. The contrast sharpened further during the Erligang period, with only one young and seven adult skeletons discovered. To further explore the age profile in various types of depositions, number of adult individuals is plotted against the number

Table 5.5 Body element distribution of each animal taxon from feature contexts at Wangchenggang

Phase	Body element	Animal taxa			
		Deer	Sheep	Cattle	Pig
Longshan	Head	1	0	2	24
	Axial	0	0	0	1
	Forelimb	0	0	1	9
	Hindlimb	1	0	1	1
	Extremity	0	0	0	1
Erlitou	Head	0	0	1	1
	Axial	0	0	0	0
	Forelimb	0	0	0	0
	Hindlimb	0	0	0	0
	Extremity	0	2	0	1
Erligang	Head	1	1	4	33
	Axial	0	0	2	5
	Forelimb	0	2	3	20
	Hindlimb	1	0	7	15
	Extremity	6	2	1	4

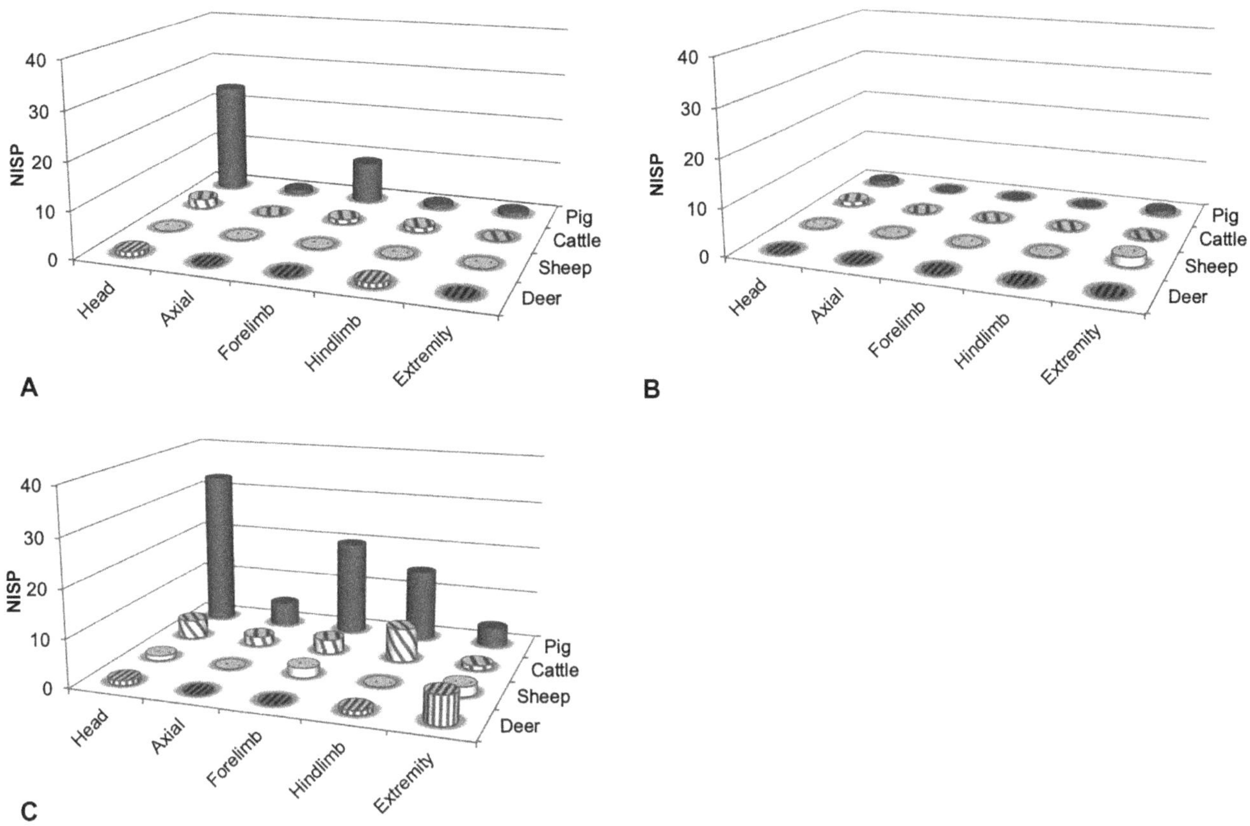

Figure 5.11 Body element distribution of major animal taxa at Wangchenggang from (A) Longshan, (B) Erlitou and (C) Erligang periods

of young individuals for each type of feature. Four general feature types are compared: foundation pits, formal tombs, pits associated with city wall or ditch and workshop contexts. As highlighted in **Figure 5.13**, the distribution of age groups is spatially uneven. Adult groups outnumber young groups in foundation pits and workshop depositions. Those features displaying a balanced distribution of the two groups contained very small numbers of individuals, usually one or two. Due to the small number of the sample for study and changing feature types through time, it is impossible to suggest a chronological trend in terms of the selection of age group associated with particular types of deposition.

5.3.7 Bone modification

In total, 21 pieces of animal skeletal specimens, including antlers, display traces of modification. An additional 11 bone fragments from unidentifiable taxa show evidence of burning. In **Table 5.8** and **Figure 5.14**, animal taxa are plotted against different types of modification, in order to investigate the different treatment of animals. Due to the small sample size, data from different periods were clumped together for analysis.

The majority of gnawed bones are pig remains, while no deer or cattle bones display evidence of gnawing. Burning and polishing are the two major types of modification occurring on unidentified bones. This is explainable, as charred bones tend to be more fragile and consequently easily fragment into unidentifiable pieces. The scenario can be interpreted in reverse: fragmented bones will more likely burn particularly when encountering natural fires. There are also burnt pig and cattle bones. Some were penetrated at the same time, probably modified as oracle bones. In terms of cutting, deer antlers have the largest proportion, followed by cattle bones. To date, no pig bone has been discovered with cut marks on it. Polishing represents the final type of modification is polishing. It is unsurprising that most bones with polishing modification belong to unidentified specimens, as surface polishing always tends to remove characteristic bone features. It is always difficult to identify specimens which have undergone thorough modification. In one scapula specimen, polishing was conducted along with

Table 5.6 Age profile of pigs at Wangchenggang (summarised from SAMPU & HPIACR 2007)

	Juvenile	Immature	Sub-adult	Adult	Elderly
Longshan	0	1	0	0	0
Erlitou	0	0	0	0	0
Erligang	10	3	0	0	0
Yingxu	0	0	0	0	0

Table 5.7 Human age profile at Wangchenggang

Period	Young	Adult
Longshan	9	15
Erlitou	0	1
Erligang	1	9
Total	10	25

penetration. This specimen was very likely prepared as an oracle bone, though heavy modification means the taxon remains unidentifiable

In terms of human remains, two Erligang skulls seemed to have chop marks, as is mentioned in the excavation report (HPIA & ADMCH 1992). The information is descriptive and is not recorded by photos.

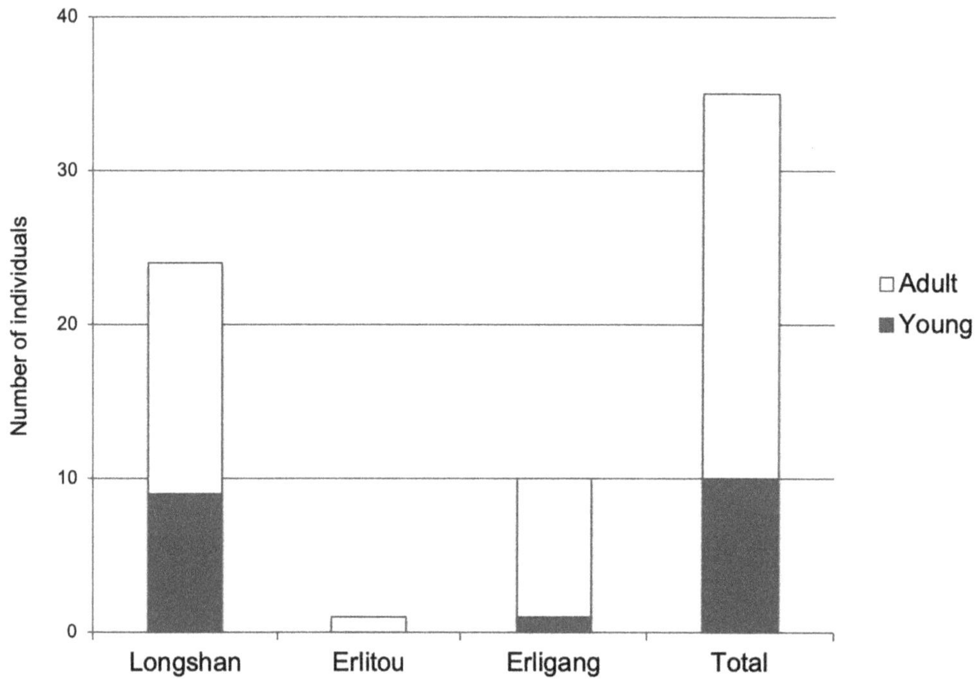

Figure 5.12 Comparison of young versus adult human groups at Wangchenggang

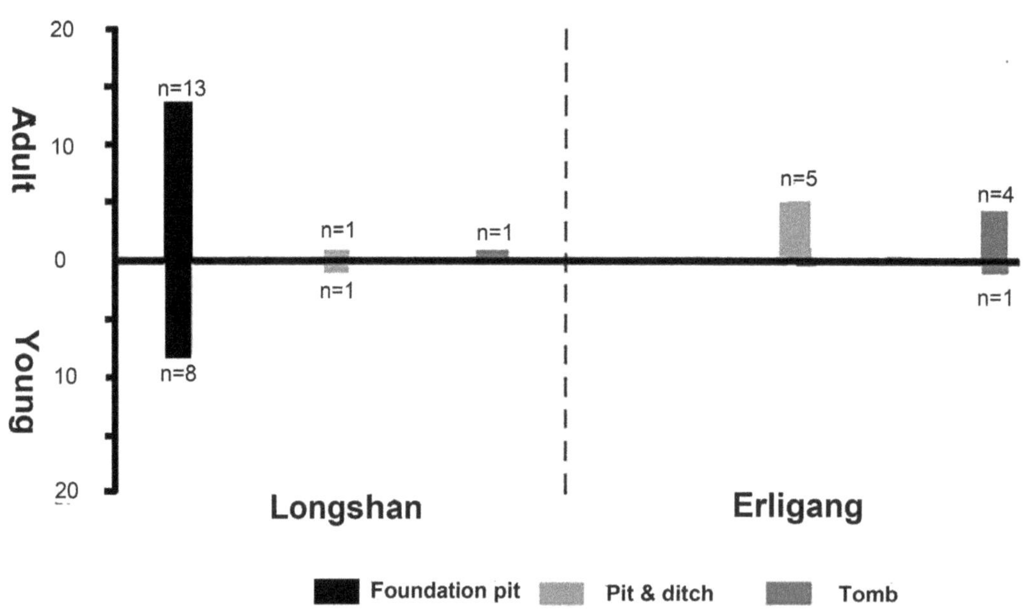

Figure 5.13 Comparison of human age profiles in different types of feature at Wangchenggang

5.3.8 Type of deposition

Inferred from the results above, three types of depositions can be identified: multi-increment, formal and refuse depositions.

Multi-skeleton-increment deposition

The first type, 'multi-skeleton-increment' deposition, refers to foundation pit 1 (**Figure 5.4**), pit 5 and pit 12 of the Longshan period, where skeletons of male, female and child skeletons were buried together; the Erligang period deer pit (WT227H555) (**Figure 5.15**) of in which two complete deer skeletons were placed, and the Erligang period surface deposition of two pigs within the house 85WT265F1 dating.

A shared characteristic of these depositions is the multiple complete or nearly complete skeletons. In addition to bones, ceramic vessels, along with a few of stone and bone tools were also excavated from this type of depositions. In the remaining foundation pits 2, 3, 4, 6, 7, 8, 9, 10, 11 & 13, either single human skeletons or incomplete body parts were discovered. Recalling the excavation strategy, all foundation pits except foundation pit 1 are yet to be fully revealed. As a consequence, many more skeletons might have been interred but have not yet been excavated. The possibility that the foundation pits mentioned above are members of multi-interment depositions cannot be totally eradicated, as in many other aspects they did resemble foundation pits 1, 5 & 12. Differences among these multi-interment depositions, one of which is a type of filling soil, do remain. While foundation pits were filled with stamped layers of rammed-earth, soil in other depositions seemed much looser and softer.

Table 5.8 Number of specimens with surface modification in different animal taxa at Wangchenggang

	Pig	Cattle	Deer	Unid
Gnawing	4	0	0	1
Burning	1	2	0	11
Cutting	0	3	4	0
Polishing	0	1	0	2
Total	5	6	4	14

Formal tomb

Another type of deposition refers to the formal tomb. These tombs were formal interments of human corpses within clearly demarcated rectangular locales along with funeral items such pottery, stone and bone tools. They are tombs 85WT270 M56 from the Longshan period, WT27 M19 (**Figure 5.16**), WT18 M43, WT15 M16, WT23 M22, WT27 M20, WT8 M15 of the Erlitou period and W2T656 M5, W2T656 M6, WT9 M8, WT19 M11, WT16 M14, WT90 M28, WT8 M6 of the Erligang period. Animal bones were only rarely buried together with these human skeletons, except for tomb W2T656 M5, where an unidentifiable animal scapula was also found though it remains uncertain whether this specimen was brought into the tomb intentionally, and WT9 M8, where a dog was deliberately placed on the right side of the human skeleton.

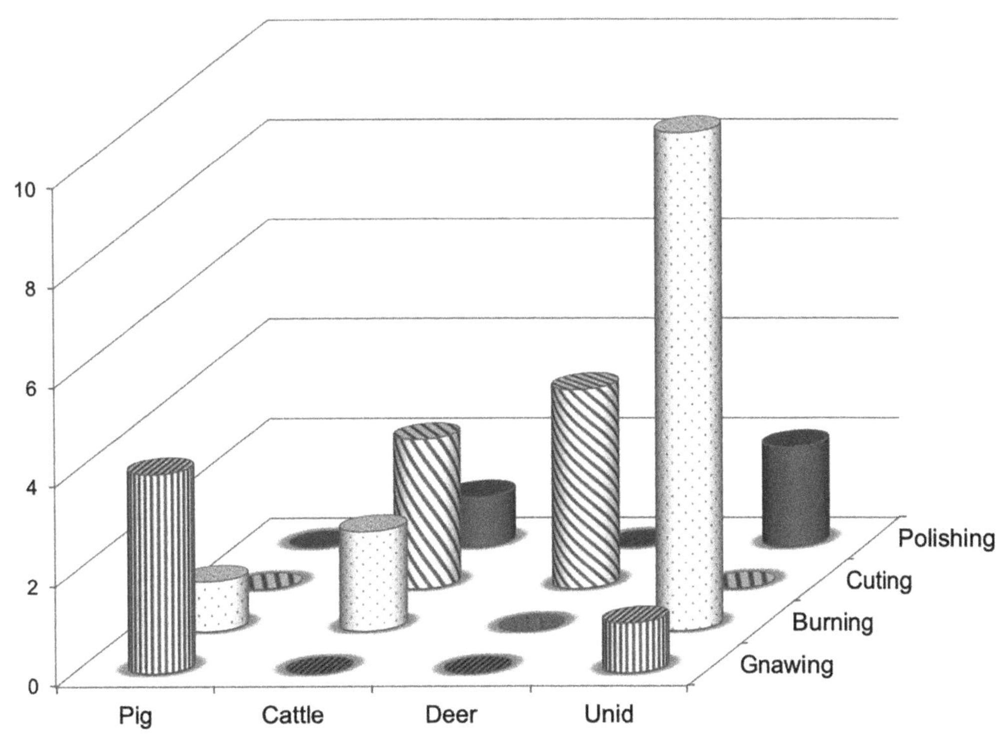

Figure 5.14 Distribution of surface modification on bones at Wangchenggang

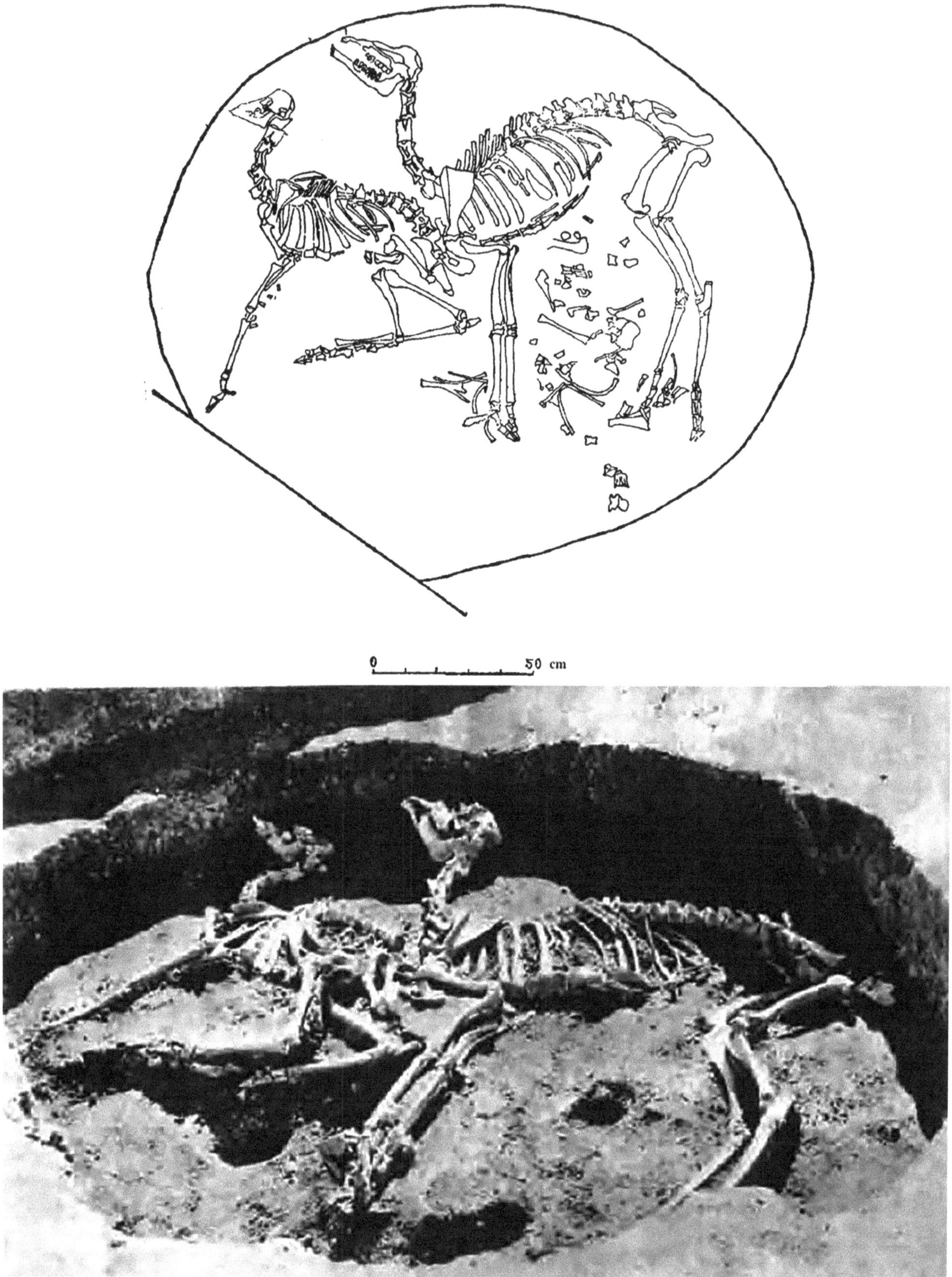

Figure 5.15 Plan drawing (upper) and photo (lower) of pit WT227H555 (deer pit) at Wangchenggang (HPICR & CHM 1992: figure 85 & photo 57-3)

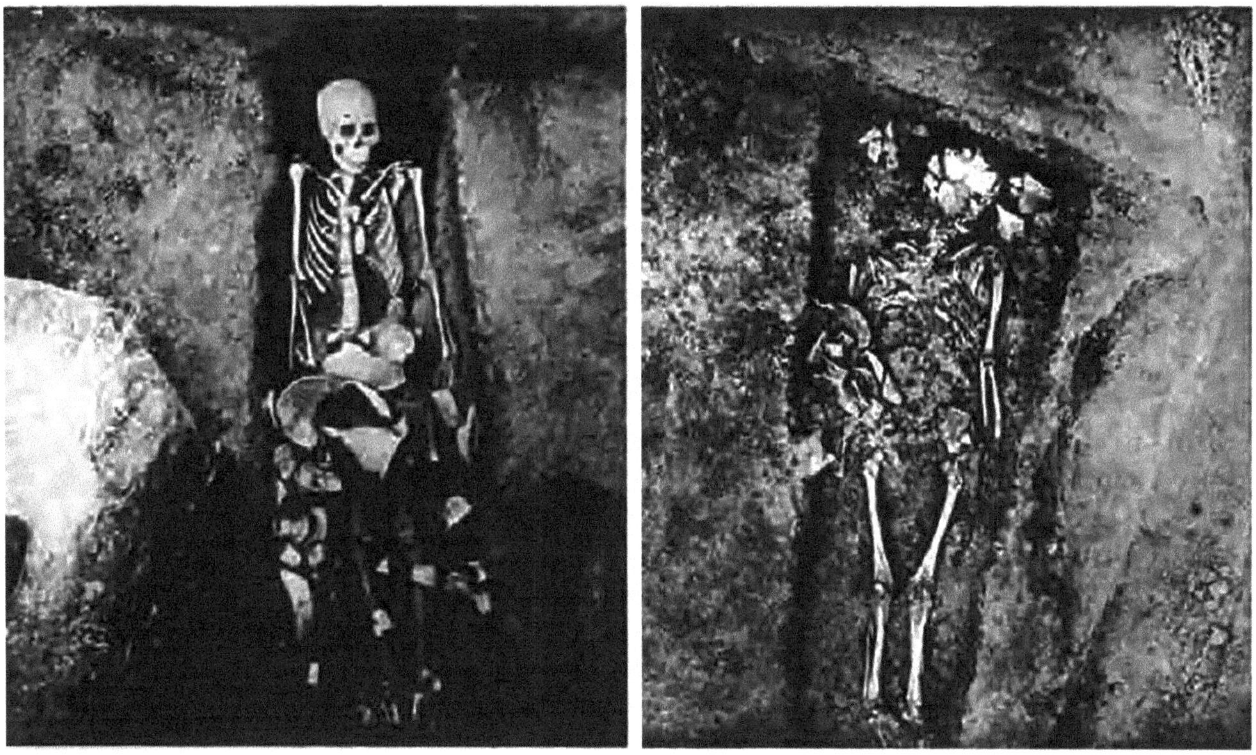

Figure 5.16 Photos of tomb WT27 M19 (left) and tomb WT15 M16 (right) from the Longshan period at Wangchenggang (HPICR & CHM 1992: photo 44-2,4)

Refuse deposition

The remaining majority is comprised of refuse depositions including those from daily consumptions and workshop waste. Both animal bones and a small number of human bones found in this type of deposition seem to be the results of random disposal, such as pit WT0672H76 of the Longshan period and pit H40, the accumulation of skull fragments in W5T0544. These depositions clearly lack deliberate arrangement after disposals which would suggest human curation. Their contents may be seen as rubbish.

5.4 Discussion

Since Wangchenggang was a massive site where excavations could hardly cover the entire settlement area, our information available is fragmented and derived from various sources. The discussion here first and foremost attempts to present a general view on animal categories on a broad scale that takes the whole site into consideration. If necessary, a magnified discussion follows, in which resolution of the dataset is permitted. The discussion begins with the comparison of treatment of different animal taxa and explores the link between animal exploitation and their categories. A continuing discussion on categories further examines the subdivision within one animal taxon. Lastly, beyond non-human animals, issues pertaining to the classification of humans are discussed at length.

5.4.1 Animal exploitation and classification

The faunal assemblage was not diverse in terms of animal taxa. The results of the analysis are consistent with the previous understanding of the lifeway at Wangchenggang as characterised by the heavy dependence upon domestic animals. This exploitation of domestic animals is believed to have roots in the Yangshao period of the middle Neolithic. At Wangchenggang, pig remained the main animal resource from the Longshan period to the Shang dynasty, while sheep and cattle gradually gained in importance during the Erligang period. This trend broadly preceded the increasingly intensive exploitation of cattle in the late Shang. The different usage between animal groups may indicate the difference in category perceptions.

In the late Shang dynasty, cattle were perceived as animals of both economic and symbolic significance, and thus they were usually treated distinctly (Campbell 2015). For example, cattle were the only animals other than turtle entitled to provide their scapula bones as the medium for divination (Keightley 1978). Faunal evidence at Wangchenggang suggests that pigs and cattle might have been treated differently. Information pertaining to bone modification enables us to demonstrate this difference. While animals appearing at Wangchenggang, such as pigs, cattle, sheep and deer provided a source of food, their involvement with human society did not conclude at the point of consumption. They were further disposed of, recycled and again engaged in human activity, probably with a renewed appearance and/

or characteristics. Referring back to **Table 5.8** and **Figure 5.14**, modification types between different animal taxa have been compared. Gnawed bones might have been exposed to the open-air without curation for a period after their abandonment and prior to their deposition. These bones were most probably food residue that had lost its human value. At Wangchenggang, gnawing occurred most frequently on pig remains, suggesting pigs were served largely as food and their bones were disposed without further curation. Once consumed, their life as useful resources was over.

Cattle demonstrate a contrasting scenario. No cattle bones were discovered bearing gnawing marks, while other types of modification including cutting, burning, polishing and drilling, occurred frequently. Unlike gnawing, these types of surface modification tend to be more suggestive of human agency, indicating possibly intentional curation and exploitation. In other words, cattle, if consumed, provided not only food, but also other resources such as raw materials for bone tools and oracle bones, once their meat had been consumed.

Taking this fact into account, it is plausible to envisage two different categories for pigs and cattle respectively, reflected in the treatment of animals post-consumption. Yet in most situations pigs and cattle both ended in refuse depositions despite differences in bone surface modification, implying that under certain circumstances pigs and cattle might have fallen into one general category denoting 'food' broadly. Consequently, classification was something more than one simple step. The classification of pigs and cattle here involved at least two levels: on one level, both were 'food'; and on the other level, both were subject to further value categories post-consumption.

It seems that distinct categories of cattle did not appear simultaneously once emerging at Wangchenggang. One reason for this may be the extremely limited volume of cattle remains at early Wangchenggang. From a broad point of view, the establishment of this category may be better regarded as a process terminating at the Shang dynasty when cattle, at which definitely imbued with special importance. The category of cattle at Wangchenggang hence was likely an intermediate state.

Deer, though being the most numerous 'wild' animals found at the site, actually rank low in NISP counts. In the contrast to the paucity of deer body remains (NISP=8), antlers (NISP=4) comprise a fair proportion of partially modified bones. Antlers occupy a rather large proportion of the deer remains. Even more antlers were discovered in layer contexts. We imagine that domesticated animals might have supplied sufficient food resources, removing the need to exploit wild animals. At the same time, instead of providing meat, deer rendered animal products such as antlers as raw materials for producing tools. The discovery of antlers both with and without pedicles indicates that people both cut antlers from corpses and at the same time collected shed antlers. So far there is too little evidence to make any conclusion about deer at Wangchenggang.

5.4.2 Age as a classificatory filter

Age might have been a theme of animal classification at Wangchenggang. This time, age as classificatory filter, served to bring different age groups together, instead of separating them.

The age filter became clearly visible during the Erligang period, evidenced by the discoveries of the deer pit WT227H555 and the pig pit H40. According to the excavation report (HPIACR & CHM 1992), pit WT227 H555 (**Figure 5.15**), with its oval opening and straight walls, was located in the south part of the site. The long axis of the opening is 2.12m and short axis 1.81m. The excavation stopped at a depth of 0.70m and the bottom of the pit had not yet been revealed. Two deer skeletons were placed at a depth of around 0.32m, one likely belonging to a large adult sika deer female , as no antler was attached, while the other belonged a smaller young individual. Both of lay on their side in a similar posture, heads facing north and faces oriented southeast. There was also a number of pottery fragments randomly scattered around the skeletons.

Pit H40, located in the house 85WT265 F1, which also dated to the Erligang period, yielded two complete pig skeletons. This pig deposition was situated in the southwest quarter of the house F1, tight against the doorway. The mature pig skeleton lied in the west with its head placed towards north and face to east, while the juvenile pig was placed a meter or so away from the mature individual. Ceramic vessels recovered within the house, but the exact locations were not reported, so it is impossible to discuss their spatial relation to pig skeletons.

In either case, it seems that one adult animal together with one juvenile animal were deliberately selected as a bounded package to be placed within the same deposition. The selection was intentionally made through two age groups which were ultimately were brought together. The phenomenon was unlikely coincidental, as it occurred with both pigs and deer, although we must admit to finding only one single example for each taxon.

We find two examples of the double depositions of one adult and one juvenile animal at Wangchenggang, with no depositions recovered containing human remains. The change in human burials caught archaeologists' attention, rendering a potential clue for understanding the role of age in classification.

For the Longshan period, no tombs for children or infants were excavated, but children seemed to be used as sacrifices probably associated with wall/ditch construction, as seen in example of H64 in W5T2372. The pit H64 was a small round pit situated adjacent to the north wall of the large enclosure (SAMPU & HPIACR 2007: 73). In the pit, an almost complete child skeleton was placed head to the north and face down with scattered pottery, stone and bone tools. No other faunal remains were excavated outside of a small number of shell fragments. These differed from formal tombs of adults in several ways, including the shape of the burial (round versus rectangular), the spatial arrangement of burial items (underneath the corpse versus around the corpse), the inclusion of shell remains or not, and so on. Therefore it might be inappropriate to regard the deposition at H64 as representing a formal human tomb. In fact before the Erligang period children had always been kept out of formal tombs. The first example of the tomb of a young individual was tomb M5 in square trench W2T656. The burial pit was cut into a rectangular shape and funeral items were included that resembled an adult tomb. The three-year-old child skeleton was placed with its head facing north and face turned west while the nearby adult tomb M6 was constructed aligned along an east-west axis (SAMPU & HPICRA 2007: 271-2).

Children are consistently missing from archaeological records. In terms of demography, the absence of child tombs must not have reflected the demographic composition of the relevant society. One reasonable explanation is that children were not buried in the same manner as adults. Archaeological data have provided ample examples in which children were buried differently from adults (Kamp 1998, 2001). It is also informed by ethnographical records that this phenomenon may be because of the distinctive perception of children from adults (Kamp 2001). For example, in many pre-industrial societies children are not viewed as 'human-beings' until they have experienced rites of passage (Eliade & Trask 1958: 132). At Wangchenggang, the absence of children's tombs during the Longshan period echoes the ethnographical generalisation about the perception of 'child' not necessarily being a member of 'human' group.

At the same time, children might have carried symbolic importance as ritual offerings. Plenty of archaeological data worldwide have recorded use of infants and children as sacrifices (*e.g.* Broda 1991; Sanchez 1993; Sillar 1994). Children were also recovered deposited in pits next to defensive walls at Wangchenggang, having probably served a sacrificial purpose, as suggested by the excavators (SAMPU & HPIACR 2007). In the Erligang period, when the earliest formal tomb of a child appears, the double deposition including one adult and one juvenile animal happened to occur simultaneously. Coincidentally, both events are more or less age-related and may be further associated with age as the classificatory filter.

As we saw at Wadian, where age categories were also constructed in the classification of animals, it seems that age was always a major theme in such classification. There might have been some changes in evaluating age-related elements in the ideological domain. Moreover there were different arrangements of age categories, 'combination' at Wangchenggang, and 'separation' at Wadian. No matter whether it served to separate or combine groups, there seemed little doubt that age was one vital factor in classifying animals.

5.4.3 Subgroups of human being

Clearly, not all human beings were treated equally. Roughly three subgroups of humans appeared at Wangchenggang: those buried in decent tombs, those deposited in pits in the form of almost complete skeletons, and those whose fragmental body elements were accumulated in refuse assemblages.

Humans buried in proper tombs were largely adults, a phenomenon that remained unchanged throughout the occupational history of the site. They were generally buried together with a certain volume of grave goods.

At first glance the age and gender of these individuals is quite mixed. Seemingly, biological characteristics were not the primary reasons behind their presence in this pit type, such as foundation pit 1 within the small enclosure. The factor that brought these skeletons into the pits seems to be invisible in bone morphology and taphonomy. The Longshan period is generally considered as an era of frequent violence and warfare (Underhill 1998). Some archaeologists associate such types of human depositions with war captives (Wang 1982; Wang 1999). Numerous records from the Shang dynasty and onwards demonstrate that captives of wars and 'barbarians' on the edge of the Shang realm were treated as inferior to citizens as quasi- or semi- human beings overlapping with non-human animals (Fiskesjo 2001 & 2012). Possible evidence for violence at Wangchenggang is derived from the skulls with chop marks, belonging to human bones which ended up in refuse depositions.

The third type of human depositions contained mainly bone elements, together with pottery fragments, broken animal bones and other pieces of artifacts. These were randomly scrambled and very likely rubbish dumps. This type of human remains was relatively rare compared with the previous two, and has only been found in three pit contexts (W5T05444 H13 & H21 of Erligang period, W5T0670 H74 of Longshan period). It remains unclear whether these human bones were deliberately disposed as trash. More analyses, for instance, isotope analysis casting light upon their difference in chemical composition of bones in different types of deposition, may provide insights into this issue.

The above three ways of treating human bones tentatively indicate that certain groups of human were perceived as a category distinct from non-human animals. Corpses may have been specially curated after death. Groups of human appearing not to belong to 'human' category deserving proper tombs can also be found, as instanced by non- or semi-human the potential 'captive' bones. The category of 'human being' thus was not merely determined by biological traits. Though invisible archaeologically, the definition of 'human' category might have involved criteria beyond biology.

At Wangchenggang, age seems to be one criterion considered when determining the inclusion of the 'human' category. Young individuals who had not reached 'adulthood' might have been excluded from the 'human' category. This scenario at Wangchengang in fact is still valid in modern times worldwide, where infants, children and teens are considered not fully developed physically and mentally under certain situations (*e.g.* drinking, voting) excluded from adult groups. In addition to age, we may plausibly speculate more criteria working for classifying humans, being it origin, religion, wealth, health, to name just a few. Put otherwise, social identity seems to be vital fact for unraveling the category of 'human'.

5.5 Summary of animal categories at Wangchenggang

The Wangchenggang site was a massive settlement bestriding the Neolithic through dynastic period. Two characteristic Longshan walled enclosures suggest its pivotal role in the contemporary regional network. Although walls and ditches were abandoned during the Shang dynasty, Wangchenggang apparently remained a considerable village.

The exploitation of animal resources was heavily dependant upon livestock. Pigs were always the main supplier of animal products. With the introduction and intensifying utilisation of cattle and sheep/goat towards the Shang dynasty, cattle very likely formed a distinct category from other livestock such as pigs, as manifested in the different treatment of bones. This tendency broadly agreed with the increasing economic and symbolic importance of cattle during the late Shang.

Age was one categorical filter at the site. The role of age as classificatory criterion did not occur only at Wangchenggang, but can also be found at Wadian, as discussed in chapter four. Young individuals were distinguished from adult individuals. Unlike the separation between the young and the adult at Wadian, residents at Wangchenggang alternatively bounded these two groups and presented them together, as seen in double animal depositions. Whatever the method of representation – separation or association – age was always the main theme. An age filter consistently modified the human versus non-human classification at Wangchenggang as well. It was not only animals that could be non-human: it is very likely that young humans did not qualify as 'human' before adulthood. That is to say, it was the social identity, rather than biological traits, that defined the category of 'human' at Wangchenggang.

6

Animal Classification at Xinzhai

This chapter examines and discusses skeletal remains from the Xinzhai site. This large walled site is believed to be an important settlement dating to the transitional period between prehistoric times and the first confirmed dynasty in Chinese history — the Shang dynasty. We begin with an introduction of the site, followed by a critical review of previous faunal studies. The results presented are primarily data from one of the most recent excavation reports. In the discussion section, issues concerning animal husbandry, exploitation of wild resources, age-related classification and the association between spaces and categories are addressed.

6.1 Xinzhai: site introduction

The Xinzhai site, located at Xinzhai village, Xinmi City, Henan Province is a significant fortified settlement in the Shuangji River valley, which spans chronologically from the Late Neolithic to the incipient Bronze Age (**Figure 6.1**).

The Xinzhai site sits close to the Shuangji River. Its scale covers four modern-day villages. One loess gully ran across the site from the east to the west while, there is another gully ran from the north to the south in the northeast . The north edge of the site is the Shuangji River (**Figure 6.2**). The site was first discovered in 1964, when the curator of the Xinmi County Cultural Museum directed an archaeological survey. In March of 1979, Zhao Zhiquan, from CASS Beijing, conducted a brief excavation and reported the significant discovery of the transitional 'Xinzhai period' believed to have bridged the Longshan and Erlitou periods (Zhao 1981). This does, however, remain a controversial transitional period, one that has been intensely debated for decades. It was not until 1999 that another excavation was conducted to re-examine preliminary conclusions drawn from the earlier discovery. Two excavation areas covered both the northern and southern parts of the site: one excavation trench and two squares in the south and the north respectively. The excavation covered an area of 161.61m^2 in total and exposed over 200 pits. In 2000, another excavation was conducted as part of the 'Origins of Chinese Civilisation' project. Excavation squares were arranged in association with the past excavation while a grid composed of five squares was dug on the west terrace off the Shuangjihe River in the north of the site. In total there was an area of approximately 324m^2 excavated and over 100 pits, along with several houses recovered. Excavations were also conducted in 2002 and 2003, but the data for these has not been fully published. To date the 2008 excavation report provides the richest available information for the 1999 and

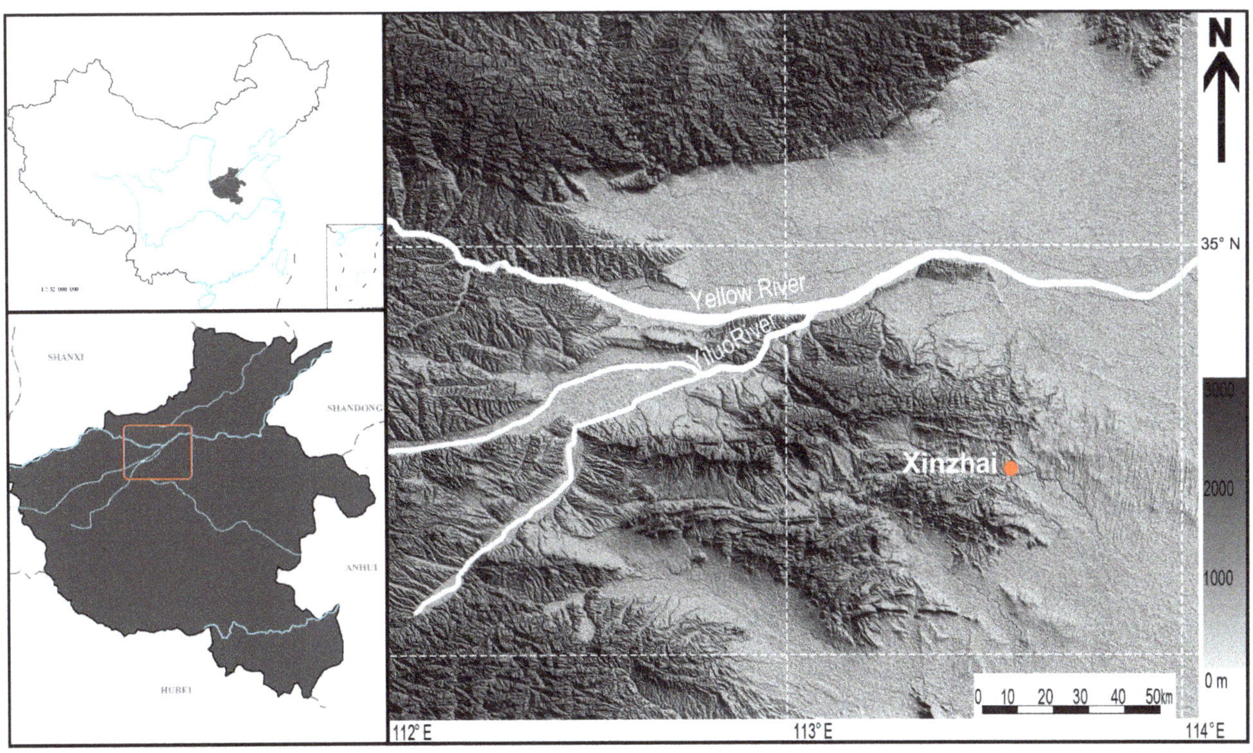

Figure 6.1 Map showing the location of the Xinzhai site

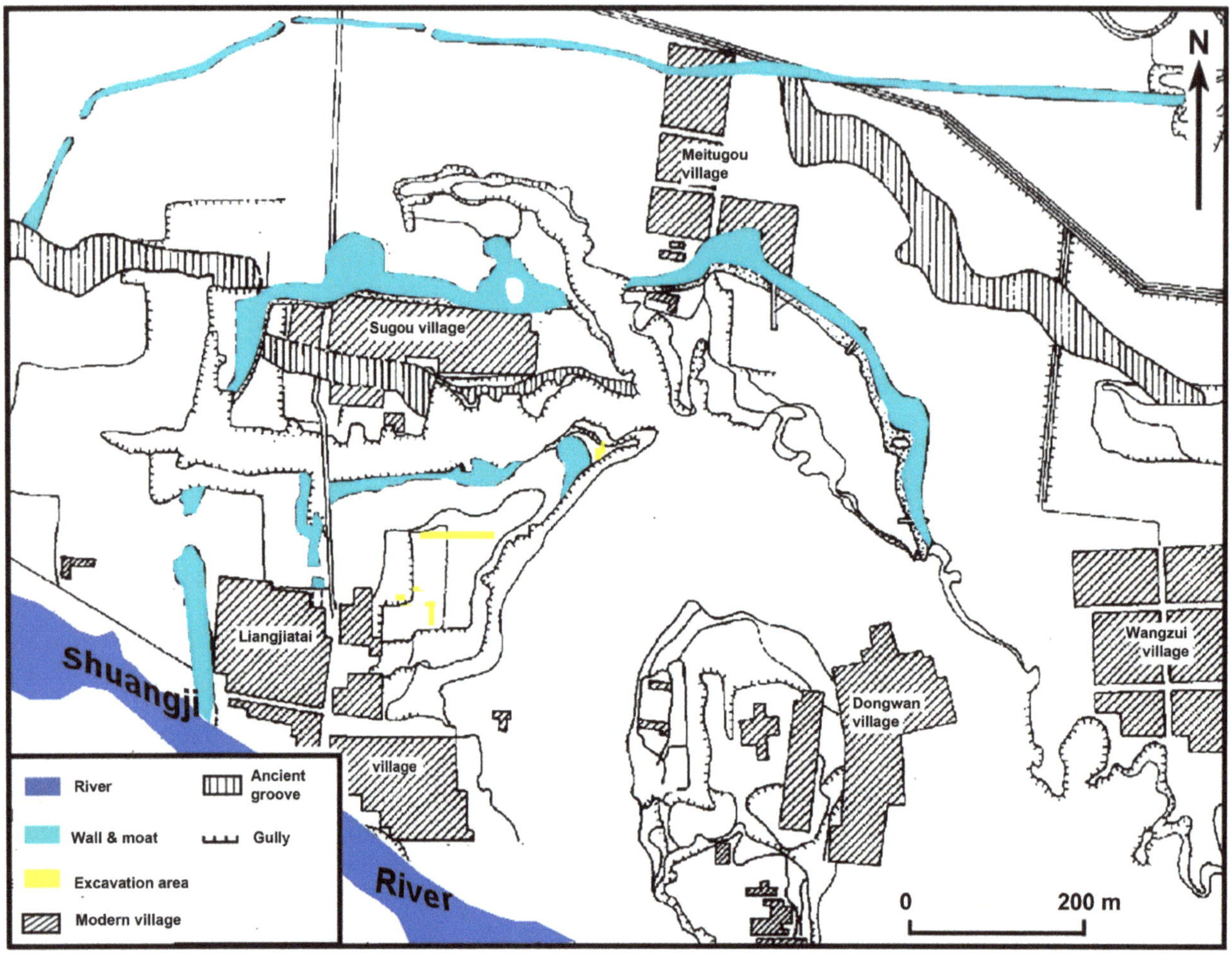

Figure 6.2 Schematic map of the Xinzhai site, showing the layout of settlement walls and excavation areas (modified from ACSACPU & ZMIACR 2008: figure 6 & Zhao 2009: figure 1)

2000 excavation seasons. The following introduction is summarised from this site report (ACSACPU & ZMIACR 2008) if not cited otherwise.

6.1.1 Chronology and phasing

All features and assemblages excavated so far are grouped into three phases, primarily based upon ceramic typology. A total of eighteen samples were collected for radiocarbon dating as well. Results are summarised in **Table 6.1**.

The first phase corresponded to the late Longshan. It coincides exactly within the 'Wangwan III Phase', the regional Henan late Longshan culture. Radiocaron dating renders an earliest date at approximately 2070~1950 BC and the latest date at around 1960~1885 BC. Taking the regional cultural sequence into consideration, the Longshan period at Xinzhai site probably no later than 1880 to 1850 BC. The pit is the most numerous feature type found for this period. In addition to pits, a small number of grooves and burials were also recovered from the excavation.

The second phase — Xinzhai period — is considered as a transitional phase linking the late Neolithic and the early Bronze Age. Feature types discovered during the Xinzhai phase include house structures, pits, grooves and burials. The preservation of house structures is very poor.

It has been this transitional Xinzhai period that has been a particular focus of debate over the last three decades. With Xinzhai initially proposed as the first period of Erlitou culture (potentially associated with the Xia dynasty within the cultural historical framework of Chinese archaeology), it was sometimes criticised as being the sole discovery of its kind (Zhao & Zhang 2010). In spite of scarce evidence, several typical Longshan and Erlitou ceramics were included in the Xinzhai culture as part of the initial proposal (Zhao 1985). Some twenty years later the radiocarbon dating results of the 'Xia-Shang-Zhou Chronological Project' challenged the first phase of the Erlitou culture as the earliest stage of the Xia dynasty, bringing attention back to Xinzhai. Thanks to the 1999 and 2000 excavations, multiple lines of evidence including ceramic typology and radiocarbon dating confirmed the Xinzhai period as the transitional period between the Longshan and Erlitou cultures, though with a limited distribution area restricted to the Songshan foothills and Yiluo River valley (Institution of Archaeology, CASS 2003: 51-52; ACSACPU & ZMIACR 2008: 531). Debate

Table 6.1 C14 dating of Xinzhai samples (summarised from ACSACPU & ZMIACR 2008: table 17 & table 32)

Phase	Lab No.	Context	C14 Dating	Cal. C14 (BC)	Cultural Period
Phase 2 Late	SA021	1999T4FH66	3425±30	1742~1695	Xinzhai period
	SA016	1999T1H29	3410±50	1745~1685	
	SA020	1999T4H30	3490±30	1770~1735(56.3%) 1710~1695(11.9%)	
	SA017	1999T1H26	3395±40	1750~1730(22.2%) 1725~1690(46.0)	
	SA013	1999T1H45	3400±55	1740~1705(63.1%) 1695~1685(5.1%)	
	SA018	1999T1H40	3500±30	1770~1735(66.1%) 1705~1695(2.1%)	
	SA010	1999T1H48	3425±35	1869~1840(15.1%) 1820~1790(18.0) 1780~1750(35.1)	
	SA009	1999TAH76	3415±35	1860~1840(15.6) 1820~1790(17.6) 1780~1750(35.0)	
Phase 2 Early	SA0028	1999T4H61(6)	3500±35	1830~1765	
	SA005-2	1999T1H112	3465±35	1825~1790(34.1) 1785~1755(29.6) 1850~1840(4.4)	
	SA0019	1999T1H115	3530±35	1830~1770(68.2)	
	SA012	1999TIH116	3480±35	1880~1845(68.2)	
	SA006	1999T1(6)C	3535±35	1884~1838(68.2)	
Phase 1 Late	SA001	1999T1H119	3485±30	1880~1846 (68.2)	Late Longshan period
	SA007	1999T1H120	3590±30	1960~1885(62.6) 1980~1965(5.6)	
	SA008	1999T1H122	3570±35	1960~1880(66.4) 2010~2000(1.8)	
Phase 1 Early	SA0014	1999T1H126	3675±35	2070~2035(63.6) 1990~1980(4.6)	
	SA002	1999T1H123	3700±65	2070~1950	

continues to rage concerning appropriate terminology for this phase (see Gu 2002; Li 2002; Pang & Gao 2008), arguing whether the assemblage found at Xinzhai site is capable of representing an independent 'culture' or a 'subtype' of the existing culture. The debate makes it clear that the transitional characteristics of the Xinzhai period are clear in both typological and chronological aspects, challenging the conventional view of the Erlitou period as the beginning of the Xia. The radiocarbon dates from Xinzhai illustrate that the Xinzhai period ranged from 1850 to 1750 BC, which fits into the chronological gap at the Erlitou site between the Longshan and Shang.

The third phase dates approximately to the early Erlitou period. No features were recovered during the 1999 and 2000 excavations, whereas only some assemblages were recognised as belonging to this period during post-excavation organisation of materials. Lacking materials for radiocarbon dating, the phase is hazarded to have begun at about 1750 BC. This date is inferred from the regional cultural sequence and site stratigraphy.

6.1.2 Spatial layout and feature types

Xinzhai was a fortified settlement protected by three sets of walls and moats along with the Shuangji River in the south (**Figure 6.2**). The outmost moats were constructed at the northern edge of the modern Meitugou village. The remaining section was roughly aligned along the east-west direction, preserved to a length of approximately 1400m. The second layer of the fortification, comprising both walls and moats, enclosed an area of approximately 7km^2. The northern wall, with the moat on its outside was recovered at the north edge of the modern Sugou village, approximately 400m north to the southern excavation area in 1999 and 2000. The west wall and moat was situated slightly to the west of Liangjiatai village, while the remnant of the east wall was situated west to the Meitugou gully. Several episodes of construction and repairs were recognised in the stratigraphy. The ancient Shuangji River turned northeast at the southeast corner of the modern Liangjiatai village. This ancient river course thus is believed to have provided a natural defense for the settlement in the south.

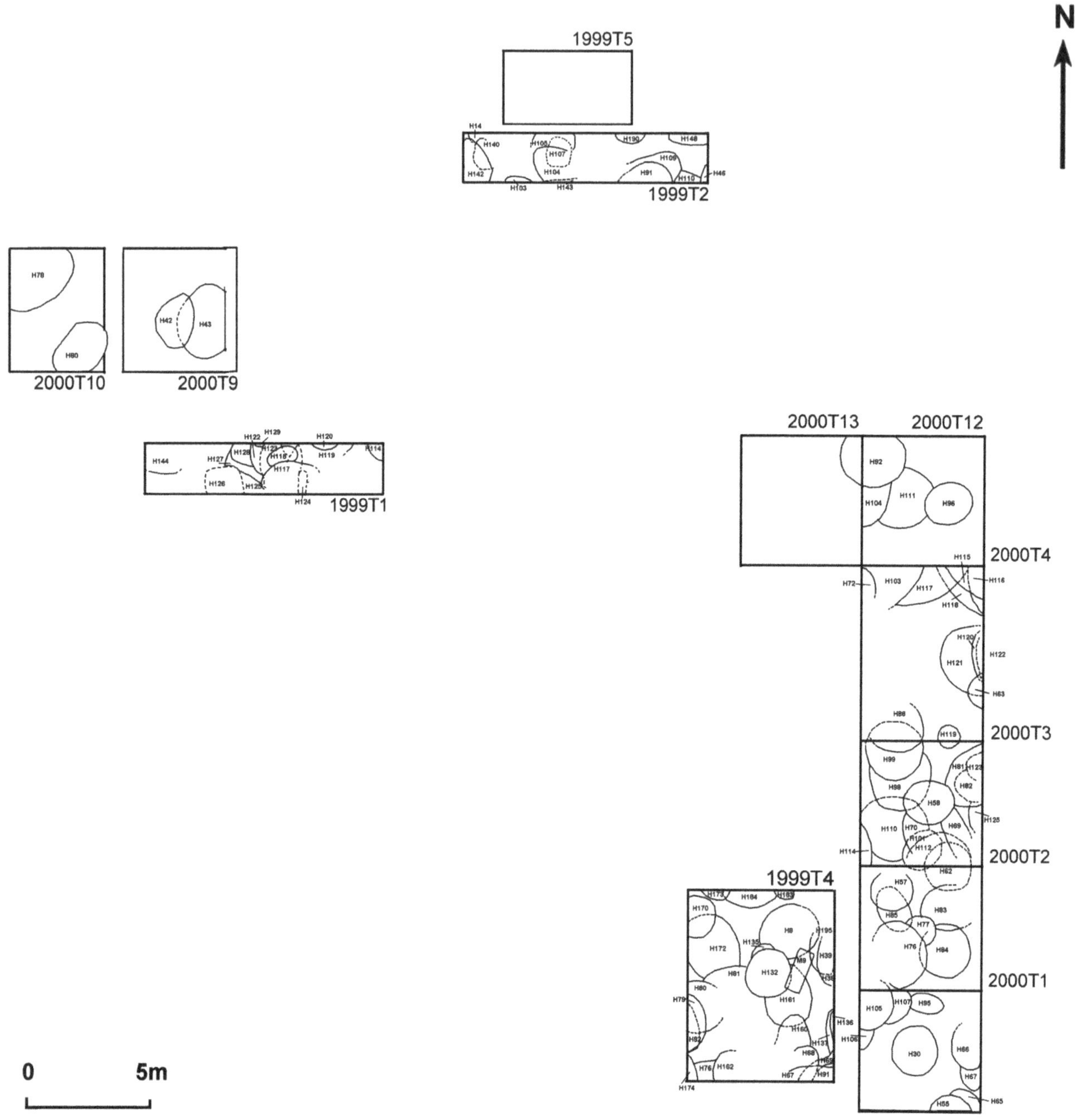

Figure 6.3 Excavation map of Xinzhai during the Longshan period (modified from ACSACPU & ZMIACR 2008: figures 15A & B)

As a consequence it is not surprising that no wall and moat was recovered in the southern margin of the settlement.

The inner wall and outer wall were located approximately several hundred meters away from the outer wall. The west inner wall was recovered along the west edge of modern Liangjiatai village. Due to poor preservation, only parts of the northern and eastern walls were recovered with severe damage.

As seen in **Figure 6.3**, pits constitute the majority of discoveries dating to the Longshan period. There were slightly fewer pits in Phase 1 than for Phase 2. These were scattered mainly in the excavation squares in the south of the site, while outside the inner ditch, far fewer pits were found in the eastern squares. Only one large structure and several house foundation, not fully fully excavated, were the sole additional finds. Thus we lack clear indications concerning settlement layout. A handful of burials was also excavated. Pits are increasingly clustered during the Xinzhai period (**Figure 6.4**). In addition to pits, discoveries of houses and burials also increased in number.

A number of house structures were reported as Xinzhai period in the 2008 report. Only three (1999T2F1, 2000T12F1 & 2000T12F2) were partially excavated while the remaining were largely disturbed as a result of feature overlap from later periods. House 1999T2F1, located in square 1999T2, was an aboveground building with five postholes. The house floor was paved with a layer of white

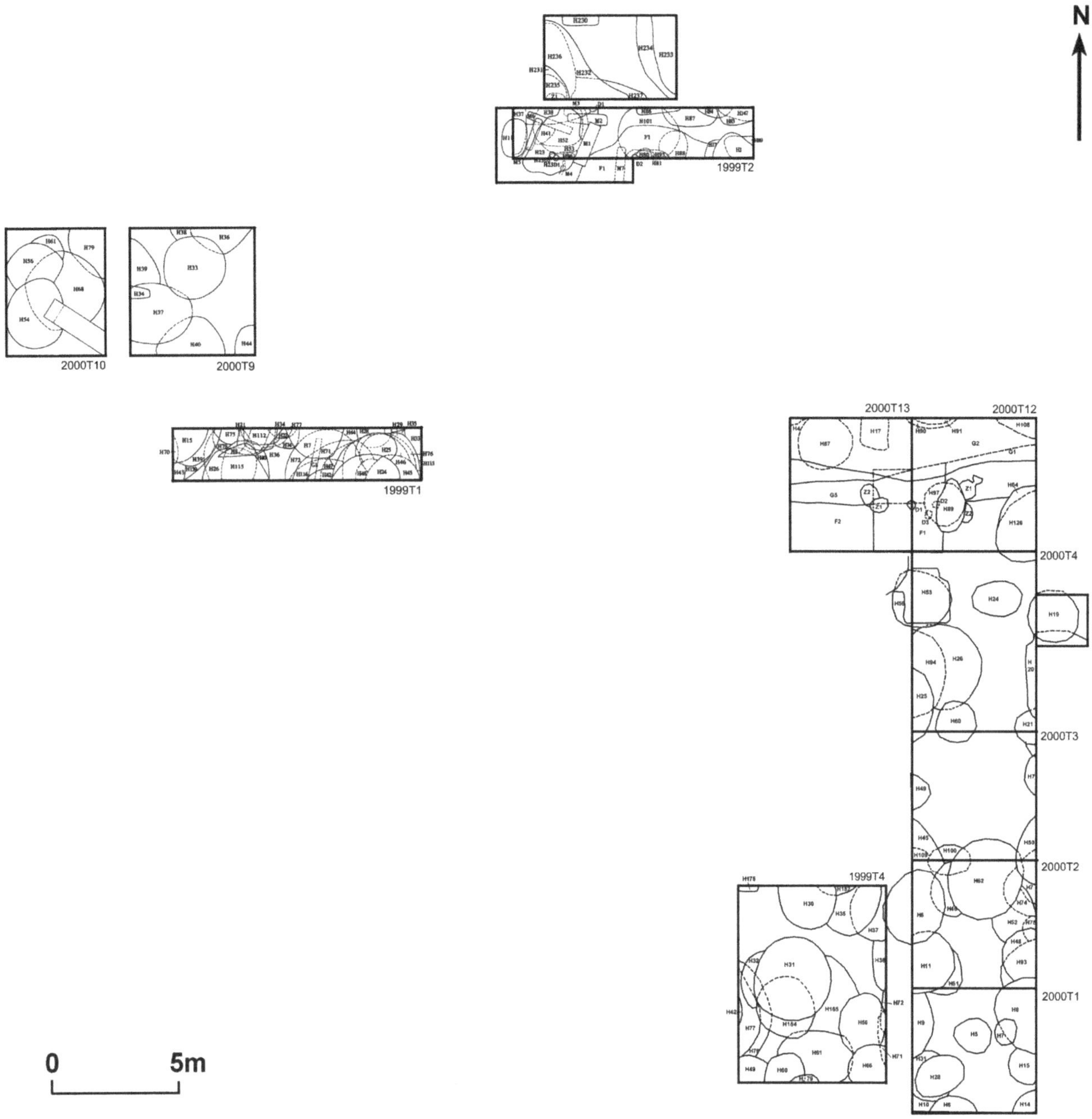

Figure 6.4 Excavation map of Xinzhai during the Xinzhai period (modified from ACSACPU & ZMIACR 2008: figures 106A, B, C & E)

plaster. Tomb 1999M7 was situated underneath this house. House 2000T12F1 was situated to the south of 1999T2F1, and contained a posthole in the middle. A total of five hearths were discovered within the house.

In addition to those normal houses, a large open-air structure was reported in another brief excavation report from 2002 (Institute of Archaeology, CASS & ZMICRA 2009). The large structure belonging to Phase 2 was found in the raised tableland in the central north of the inner city, approximately 50m south of the inner ditch (the central north excavation area in **Figure 6.2**). The structure remained 99.2m long from east to west and 14.5m wide from north to south, forming a rough rectangular shape (*ibid.*). It was dug in to a shallow rectangular pit, above which no walls and postholes were found (*ibid.*). There were several dumps of burnt earth scattered along its northern and southern edges (*ibid.*). The excavator associates this structure with a 壇 '*tan*', 墠 '*tan*' or 坎 '*kan*' , a ritual site which refers to a shallow square or pit for ritual performance, as recorded in some ancient texts (*ibid.*).

(Paragraphs above summarised from ACSACPU & ZMIACR 2008)

6.1.3 Xinzhai site and its regional network

Placing the Xinzhai site against the regional background, it can be seen that this fortified site was likely a pivotal centre in the regional settlement hierarchy. Especially during the Xinzhai period, the emerged became the only fortified settlement in this region, witnessing the remarkable shrinking of other contemporaneous sites. Many scholars argue for Xinzhai's central role regarding its settlement scale, fortified layout, elaboration of crafts, possible surplus of food and so on. (*e.g.* Ma 2004; Gao 2005; Li 2013). Some even associate Xinzhai with one of the capitals of the legendary *Qi 启* emperor of the Xia (*e.g.* Zhao 2004; Ma 2007). No matter whether once capital or not, Xinzhai might have been the centre of the regional network in the area surrounding Mount Song. This hypothesis is primarily supported by the size and fortification of the site. Doubtless a considerable amount of labour and production were required to build up a settlement to such a colossal scale. Xinzhai might have also held a central role in terms of regional social network in the regional landscape. This discussion will dwell upon studies which bring humans as active agents on to the historic and social stage rather than focusing on the site itself. How residents at Xinzhai could have led their life is a dispensable part of the settlement history. Animal resources were one important factor intertwined into the everyday life of residential life.

6.2 Previous zooarchaeological studies

Faunal remains were collected during the 1999 and 2000 excavations. Reports of earlier excavations do not mention whether faunal remains were collected. Faunal data from the excavation after 2002 have not yet been released to the public.

A preliminary analysis on the faunal assemblage collected during the 1999 and 2002 seasons is enclosed in the 2008 site report. It comprises several chapters (ACSACPU & ZMIACR 2008). The research counts the number of identified specimens (NISP), based on which the minimum number of individual (MNI) was also calculated. Age profiles of major animal taxa are included and measurements of bones are noted when necessary. According to the report, a total of 4604 pieces of non-human bones were recovered at Xinzhai, including shells, fish, birds and mammals. Mammal remains make up the largest portion, adding to 3840 pieces of identified bones in total. The major animal taxa are pig, deer, cattle, sheep and dog. Shell and fish remains were found in limited numbers. Small animals such as reptiles and small mammals were also present in the assemblage, albeit in small quantities. The presence of porcupine (*Hystris sp.*), Pere David's deer (*Elaphurus davidianus*), bamboo rat (*Rhizonmys sp.*) together with some riverine clams demonstrate a warmer climate in this region during the Late Neolithic and the beginning of the Bronze Age. Comparing the number of animals present, a fluctuation in the exploitation of wild and domesticated animals is highlighted through time. Pigs predominated in Phase 1, but dropped to under a half in Phase 3, while deer remains gradually became rich in the assemblage due to the environmental deterioration in Phase 2. The growing importance of sheep and cattle exploitation was also observed in Phases 2 and 3, evidenced by the increasing NISP counts.

(This paragraph summarised from ACSACPU & ZMIACR 2008: Ch 6)

Recently isotope analyses have been performed on both human and animal bones from the Xinzhai site. Pig isotopic signals demonstrate dietary intake skewed towards millet products (Zhang & Zhao 2015; Dai *et al.* 2015). Cattle were probably provided with fodder comprising millet, while sheep herds might have grazed around pastures with occasional provision of fodder depending on seasonal fluctuation (Dai *et al.* 2015). Deer dietary intake, strikingly, suggests a heavy reliance upon millet products, interpreted as an indicator of deer domestication (Zhang & Zhao 2015). Nonetheless, it reamains questionable whether we should use 'domestication' to describe human-deer relationship without further clarification. Dai and her colleagues (2015) suggest that fed deer were likely manipulated for the purpose of royal hunting, like scenarios recorded in later Shang oracle scripts and ancient Chinese literatures such as the *Shijing 詩經 (Books of Songs)*. Further linking the animal management strategies to the increased complexity of social and political intertwining, Cucchi and his colleagues (2016) propose that the pig management at Xinzhai reflects a rigid control imposed by the authority rather than the household-based pig domestication.

In addition to food provision, animals at Xinzhai were exploited for secondary products as well. Dai and colleagues (2014) speculate on wool exploitation through an examination of sheep age profiles. Bone tools were excavated on site. A bone-working area was potentially located on the terrace by the Shuangji River in the south margin of the site (Zhao 2004).

Human bones were analysed in a separate chapter in the site report. Information about bones is presented context by context, mostly in a descriptive manner without detailed spreadsheet recording. According to isotopic data, the human diet was comprised of a considerable proportion of millet products and rich animal proteins (ACSACPU & ZMIACR 2008).

6.3 Results

According to my calculations, a total of 3336 identified specimens can be found for the excavations from the years 1999 and 2000. Among them are 562 shell specimens and and two fish specimens, and six and 23 reptiles and bird specimens, respectively. 3210 specimens belong to mammal remains, including humans. 388 specimens are from Phase 1 whereas 2315 specimens date to Phase 2 and the remaining 507 specimens are from Phase 3, including human remains (**Table 6.2**). It is obvious that Phase 2 yielded the richest faunal remains.

6.3.1 Animal taxa

In addition to taxa frequently observed in the Neolithic sites, such as pigs (*Sus domestrica*), deer (including *Cervus nippon, Elaphurus davidianus, Hydropotes inermis*), cattle

Table 6.2 NISPs of major animal taxa at Xinzhai

	Pig	Deer	Cattle	Sheep	Other Mammals	Human	Total
Phase 1	157	54	59	13	32	73	388
Phase 2	946	460	216	227	74	392	2315
Phase 3	233	124	81	55	14	0	507
Total	1336	638	356	295	120	465	3210

(*Bos taurus*), sheep/goat (*Ovis sp./ Capra sp.*), dog (*Canis familiaris*) and so on, small mammals such as porcupine (*Hystris sp*), bamboo rat (*Rhizomys sp*), rabbit (*Lepus sp*), badger (*Meles meles*), and beasts like Asian black bear (*Selenarctos thibetanus*) were also found at Xinzhai, albeit in relatively smaller proportions. Human remains were concentrated in Phase 2. **Figure 6.5** demonstrates the difference in proportion among these various animal taxa. Each of the four major animal taxa present in the assemblage is discussed below. It would seem that, when human remains were included, the distribution of NISP percentages displays a rather different pattern from previous studies.

Pig

Pig remains comprised a relatively large proportion of the entire assemblage for all three phases. The original report indicates a decreasing trend in pig percentages from 60 per cent in Phase 1 to 51 per cent in Phase 2 and 43 per cent in Phase 3 (ACSACPU & ZMIACR 2008: 467). However, when human remains were included in the comparison, pig proportions remained under 50 per cent throughout all three phases. Compatible with the general trend, the number of pigs also peaked at Phase 2, followed by Phase 1 and Phase 3.

Deer

The volume of deer remains continually increased from Phase 1 to Phase 2. So too culminating in Phase 3 at 24.5 per cent, though the exact deer NISP in Phase 3 was much reduced over Phase 2. Deer is the only taxon at Xinzhai of continually increasing over all three phases.

Cattle

59 pieces of identified cattle bone were found in Phase 1, a number fairly close to deer figures in Phase 1. Although the number of cattle in Phase 2 increased drastically,

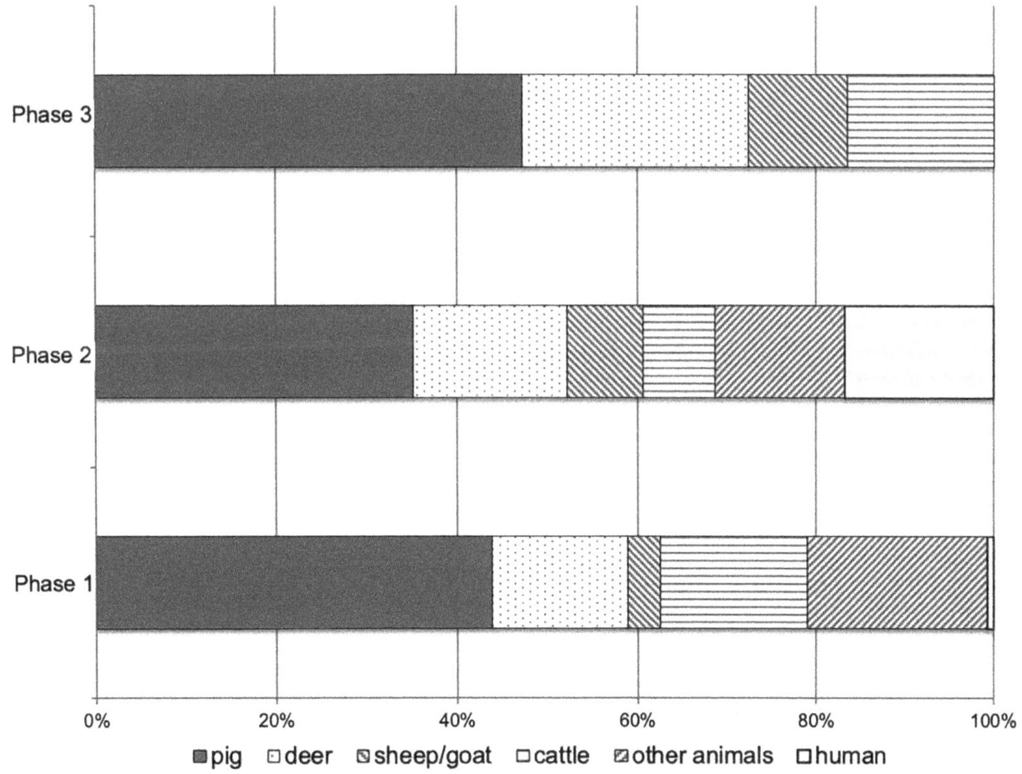

Figure 6.5 NISP percentage of major animal taxa at Xinzhai

its proportion reduced from 15.2 per cent in Phase 1 to 9.3 per cent in Phase 2. Phase 3 saw a revival in cattle proportion which reached 16.0 per cent.

Sheep and goat

Both sheep and goat horns were identified from the assemblage, indicating use of both taxa. Distinguishing between the bones of either taxa, however, remains difficult. Only a handful of sheep/goat remains were collected dating to Phase 1. The number of identified sheep/goat specimens then increased, with as many as 227 pieces of bones were identified under the category of sheep and goat in Phase 2. The shifting trend of proportion matches changing sheep/goat NISPs. Their proportions in Phase 2 is almost three-times as large as that of Phase 1, from 3.4 to 9.8 per cent and then levels out about 10.8 per cent in Phase 3.

Other non-human mammals

The category of 'other non-human mammal' includes dogs, together with some small mammals such as rabbit, badger and porcupine. Some large animals, considered to interact less frequently with humans, for example bear, are also included under this category. 30, 64 and 11 specimens of dog bones were recovered in each phase respectively. Excluding dogs, other mammals had proportions of 0.1, 1.6 and 0.8 per cent across all three phases. NISP percentages for this group were always fewer than 20 per cent.

Human

Descriptive information about human remains enclosed in the original site report was compiled into the spreadsheet in a similar style to that for faunal remains. The human NISP is 466 in total from Xinzhai among which two were excavated from Phase 1 and none from Phase 3. As a result, the proportion of human NISP in Phase 2 reached approximately 19.5 per cent, outnumbering other animals, with the exception of pig. These specimens (NISP=394) comprise 12 individuals, with most skeletal elements were preserved in 1999T2M1-9, 2000T12H96, 2000T4H53 and 2000T4H64 respectively. The remaining bone fragments (NISP=72) were deposited in pits, ditches and layers.

6.3.2 Body element distribution

Distributions of skeletal elements of four animal taxa (pig, deer, cattle, sheep/goat) in each phase, also including human remains, are summarised in **Table 6.3** and plotted in bar charts **Figure 6.6**. The comparison is not applied to Phase 3, as no human remains were recovered during this period. However, we note that over 70 per cent of human remains were preserved as parts of almost fully intact skeletons. Comparisons of deposits are also made between layer and feature contexts.

In feature deposits belonging to Phase 1, pig remains were predominated by head elements, while in other animals, by contrast, scarcely any head elements made the contribution to NISPs. There were a few limb bones from pig, deer and cattle. The body part distribution in archaeological layers of Phase 1 illustrates a different pattern from those in Phase 1 features. Bones of heads remain numerous in pigs, whereas proportions of head parts in deer, cattle and sheep are intermediate. These differences were statistically evaluated by Wilcoxon test, the results of which suggesting differences to various degrees (**Table 6.4**, **Table 6.5**). In feature depositions, among different animal taxa, pig and

Table 6.3 Body element distribution of skeletal remains at Xinzhai

		Feature contexts					Layer contexts				
		Human	Pig	Deer	Cattle	Sheep	Human	Pig	Deer	Cattle	Sheep
Phase 1	Head	2	24	0	1	1	0	11	3	2	2
	Axial	0	11	2	5	0	0	3	1	1	0
	Forelimb	0	9	8	4	0	0	6	0	1	0
	Hindlimb	0	1	1	3	0	0	2	0	0	0
	Extremity	0	1	7	1	1	0	1	2	1	1
Phase 2	Head	146	173	28	10	21	11	134	23	27	19
	Axial	140	83	32	23	5	5	51	32	24	9
	Forelimb	45	87	50	11	12	4	132	53	30	45
	Hindlimb	59	71	26	17	11	8	111	59	13	40
	Extremity	44	19	44	19	11	3	25	31	23	19
Phase 3	Head	0	0	0	0	0	0	88	22	10	12
	Axial	0	0	0	0	0	0	24	3	8	2
	Forelimb	0	0	0	0	0	0	47	18	8	16
	Hindlimb	0	0	0	0	0	0	42	21	18	9
	Extremity	0	0	0	0	0	0	24	24	28	13

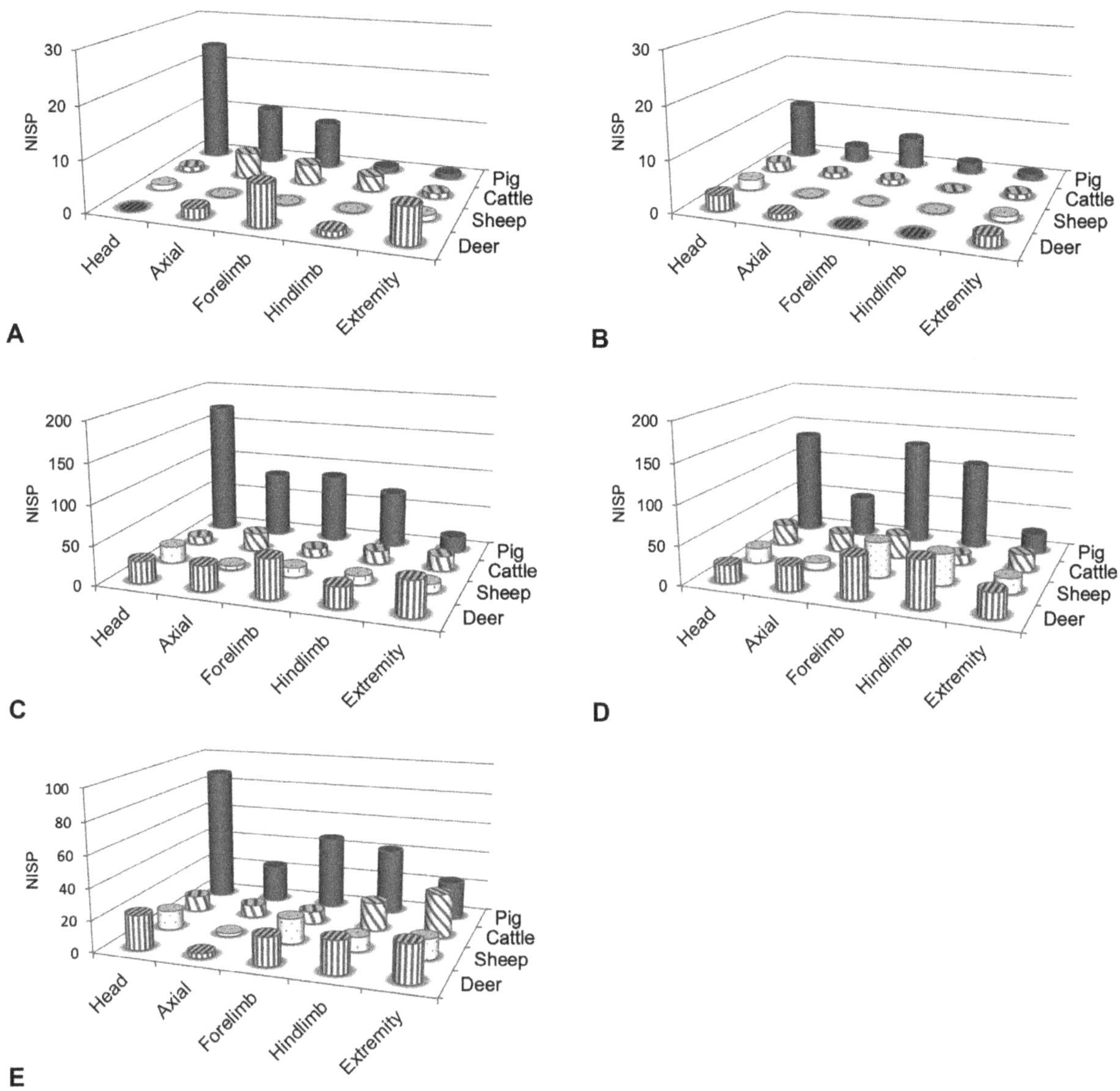

Figure 6.6 Body element distribution of major animal taxa at Xinzhai in (A) pit contexts from Phase 1, (B) layer contexts from Phase 1, (C) pit contexts from Phase 2, (D) layer contexts from Phase 2 and (E) layer contexts from Phase 3

Table 6.4 P (same) values from Wilcoxon test to evaluate the difference of body element distribution between phases at Xinzhai

Period	Sheep	Cattle	Deer	Pig
Phase 1	0.317	0.141	0.273	0.068
Phase 2	0.138	0.225	1	0.686
Phase 3	na	na	na	na

sheep body-parts are least likely to distribute similarly (p=0.068) while cattle and deer body-part representations demonstrate similarities to a certain extent (p=0.686). In feature contexts, body-part representations in cattle and deer share the greatest similarity (p=0.686) while those of pig and sheep demonstrate diversity (p=0.068) compared with other taxa. Two pieces of human bones (skull fragments) were recovered from Phase 1 features. Cattle and deer distribution shows higher possibility of

Table 6.5 P (same) values from Wilcoxon test, to evaluate the difference in body element distribution between animal taxa at Xinzhai

Phase	Context	Sheep Vs Cattle	Sheep Vs Deer	Sheep Vs Pig	Cattle Vs Deer	Cattle Vs Pig	Pig Vs Deer
1	Pit	0.083	0.104	0.068	0.686	0.109	0.225
	Layer	0.157	0.083	0.068	0.655	0.066	0.104
2	Pit	0.500	0.043	0.043	0.043	0.068	0.080
	Layer	0.686	0.686	0.043	0.123	0.043	0.080
3	Layer	0.416	0.104	0.043	0.498	0.080	0.068

similarity in layers (p=0.655) whereas cattle and pig display differences in body-part representation (p=0.066).

In Phase 2, head and limb elements made great contributions to pig assemblage in layers. In deer, the most numerous parts were from hindlimbs, followed by forelimb, extremity, axial and head elements. The body parts in cattle assemblage seem balanced comparing with other animals, while element distribution in sheep/goat appear similar to those in deer where limb bones dominate the assemblage. Statistically, patterns in sheep and cattle (p=0.686), and sheep and deer (p=0.686) bear more resemblances than other taxa, while significant differences are found between sheep and pig (p=0.043), and cattle and deer (p=0.043). In terms of pits depositions, head bones outnumber other parts in pig assemblages, although limb bones also make up a fair proportion. The number of extremity elements increases for the deer assemblage, ranking second in terms of abundance. There were still more limb bones than head and axial elements. In cattle remains, axial elements relatively outnumber other body parts. The least are those head and forelimb elements, while hindlimb and extremity elements rank intermediately. Sheep assemblage in features display a pattern distinct from layers where head elements instead became the most abundant body part. Inferred from statistical comparisons, resemblance is seldom found between animal taxa (most p values are less than 1; see **Table 6.4**), except in comparisons between sheep and cattle (p=0.500). No animal taxa displays significant difference in body-part distribution between layers and pits. Deer, in particular, have almost identical distribution in layers and pits. Human bones were present in both feature and layer contexts, the majority of which (NISP=438) were concentrated in feature depositions (*i.e.* tombs). Head and axial bones have the largest proportion. The large number in head parts in feature contexts may be attributed to the concentration of skull fragments. Only a small number of human bones (NISP=31) were excavated from layer contexts.

In Phase 3, head elements are the most abundant in pigs, just like body part representations in the previous two phases.

Table 6.6 Age profile of pigs estimated by mandibles at Xinzhai

	Phase 1	Phase 2	Phase 3	Total
Juvenile	4	7	2	13
Immature	5	14	2	21
Sub-adult	30	48	14	92
Adult	0	3	1	4
Elderly	0	1	0	1

Limb elements rank the second, followed by extremity and axial elements. Axial bones were scarcely found for deer. The most abundant elements are bones from the hindlimb and extremities, which slightly outnumbered bones from the head and forelimb. The pattern of cattle body part distribution is roughly similar to deer, albeit differing in numbers. Bones from extremities are the most numerous, followed by those of hindlimb, head, forelimb and axial elements. In sheep assemblage, forelimb elements produce the highest counts, followed by extremity, head, hindlimb and axial elements. Sheep and pig tend to be significantly different in the body-part representation (p=0.043) while those between cattle and deer share a relatively higher level of similarity (p=0.498).

6.3.3 Age profile

Age structures of pig and deer were estimated.

The age of death in pigs estimated by eruption and wearing in mandible teeth suggests these animals were killed while still sub-adults. In all the three phases, the sub-adult pig group around one and two years old always had the largest number in assemblages, followed by the immature group (**Table 6.6** and **Figure 6.7**).

The age of epiphyses fusion is presented in **Table 6.7** and plotted in **Figure 6.8**. The evidence also points to the preferred killing of pigs at a young age, consistent with estimation by mandibles.

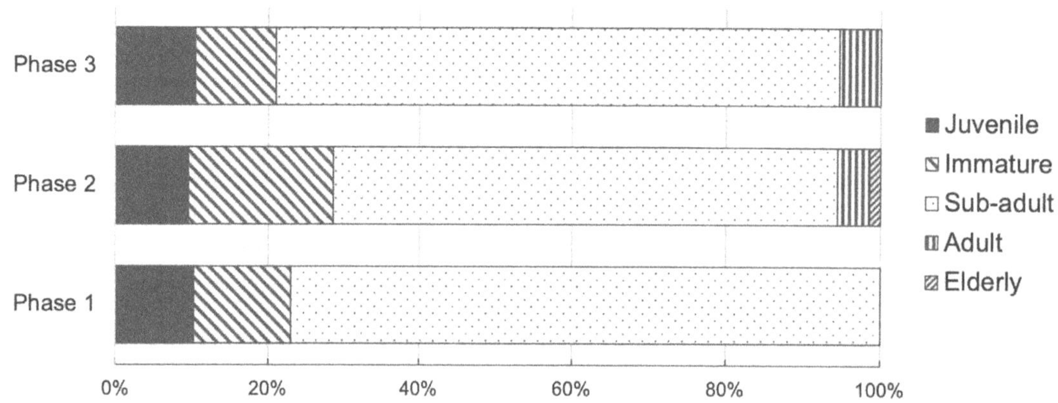

Figure 6.7 Age group percentage of pigs at Xinzha

Table 6.7 Epiphyseal fusion information in pigs at Xinzhai

	Early fusion		Middle fusion		Late fusion	
	Unfused	Fused	Unfused	Fused	Unfused	Fused
Phase 1	3	6	5	0	10	11
Phase 2	15	121	18	22	117	75
Phase 3	2	37	8	16	21	38
Total	20	164	31	38	148	124

Due to the lack of detailed records for these mandibles, the age at death in deer has been estimated into two broad groups: those with fully erupted teeth and those without fully erupted teeth. Only one deer mandible was found in Phase 1, albeit lacking information on age, and with no examples from Phase 2. In Phase 3, 21 deer mandibles were recovered, belonging to at least 13 individuals, and could be approximately aged. As presented in **Table 6.8**, about 70 per cent of deer were killed at the age of more than three years, while only one mandible specimen from 2000T1(8) layer belonged to a fawn.

Due to the small sample size of the deer assemblage for age estimation, remains from all phases are lumped together for calculation. Fusion information on deer is summarised in **Table 6.9** and plotted in **Figure 6.9**. Most deer were slaughtered after the majority of their bones had fused,

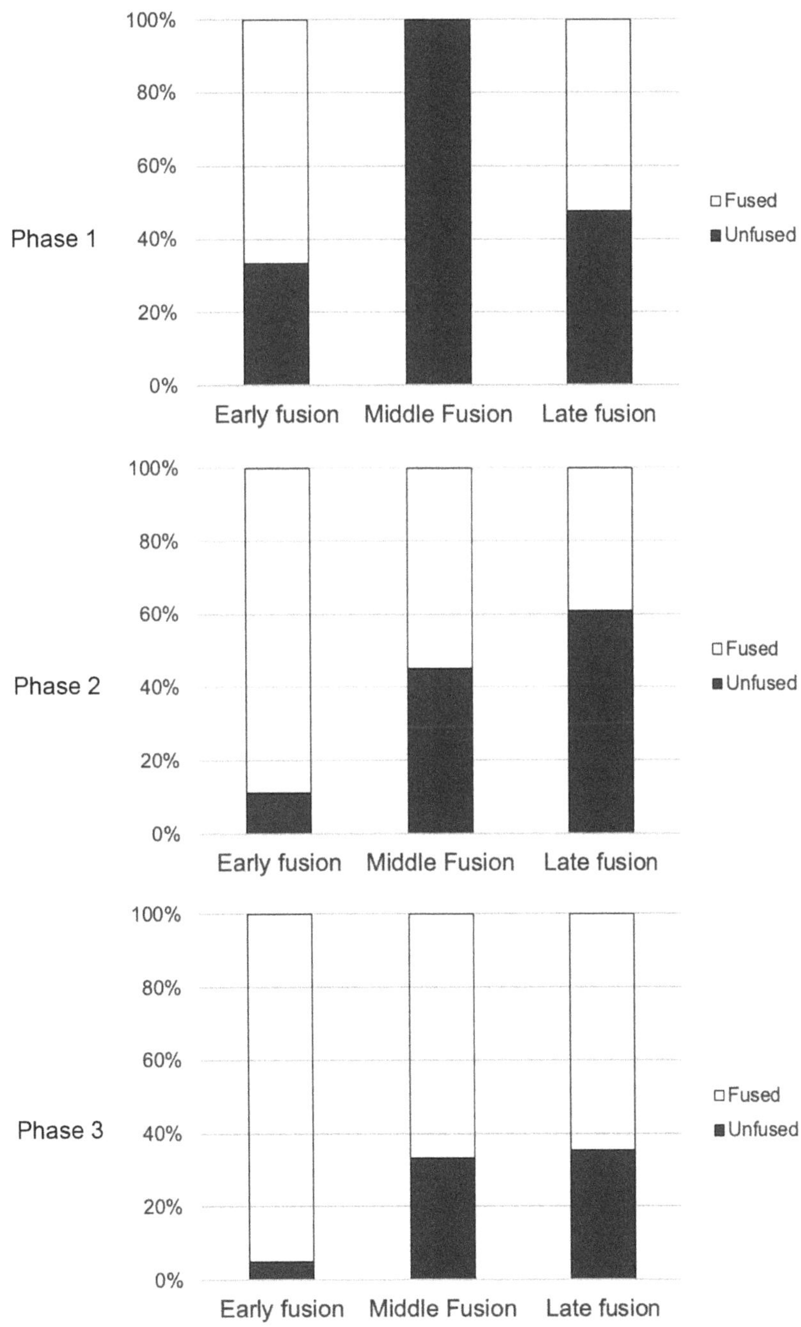

Figure 6.8 Epiphyseal fusion pattern in pigs from Xinzhai

Table 6.8 Age profile of deer estimated by eruption of mandible teeth at Xinzhai

Eruption	Age (months)	Number of mandibles	Percentage (%)
All teeth erupted	Under 24	2	28.6
Teeth unerupted	Over 30	5	71.4
Unknown		6	

Table 6.9 Deer NISPs in each age groups estimated by epiphyseal fusion information at Xinzhai

	Early fusion		Middle fusion		Late fusion	
	Unfused	Fused	Unfused	Fused	Unfused	Fused
Total	0	9	1	38	4	77

suggesting a preferred killing for matured individuals. The pattern resonates the result inferred from tooth eruption on the mandibles.

6.3.4 Associated bone group

Associated bone groups were occasionally recovered among non-human animal remains. So far only three ABGs have been found, each containing a complete pig skeleton and all dating to Phase 2. Pit 2000T1H8 was located in the northof the excavation area. Only partly excavated, its edge extended beyond the excavation area. The round opening of the exposed part was 2.95m long and 1.79m wide, whereas the bottom was 2.58m long and 2.5m wide. The depth was around 2.2m and the wall was straight. Enclosed was a complete skeleton of a one year-old sow. Child bones were deposited in the same context. A relatively large volume of ceramic vessels were excavated along with bone and stone tools as recorded in the original site report (this paragraph summarised from ACSACPU & ZMIACR 2008).

Pit 2000T2H48 was located next to pit 2000T1H8. It has not been fully excavated. Inferred from the exposed section, it was likely a round pit smaller than H8. It cut into another pit H93. A bone chisel was reported recovered from the pit (ACSACPU & ZMIACR 2008). The pig skeleton inside was destroyed, making it impossible precisely age and sex. There were human and other animal bones mixed in the pit filling.

The third pig ABG was found in the layer context 2000T7(8)A. This excavation square was located outside of the inner enclosure. The same layer contained findings of other animal bones, including deer, cattle and sheep.

In terms of human remains, recovered remains comprise of 12 individuals (NISP=394) with skeletal material for the most part preserved. It seems appropriate to label these 12 groups of human remains associated bone groups (ABGs). All ABGs were preserved in 1999T2M1-9, 2000T12H96, 2000T4H53 and 2000T4H64 respectively. The 12 ABGs were recovered from two deposition types: eight rectangular tombs and four pits with round or oval openings. One ABG comprising 73 NISP is dated to Phase 1 and the remaining human bones were recovered from Phase 2.

6.3.5 Type of deposition

Difference in deposit type is first examined by comparing depositions in feature and layer context, since it is believed that depositions in feature contexts are intentionally guided by taxonomy, while those in layers formed more or less accidentally and randomly.

Layer and feature depositions

Animal remains were collected from both layer and feature contexts. NISPs of four major animal taxa of interest were calculated and compared to test the depositional difference

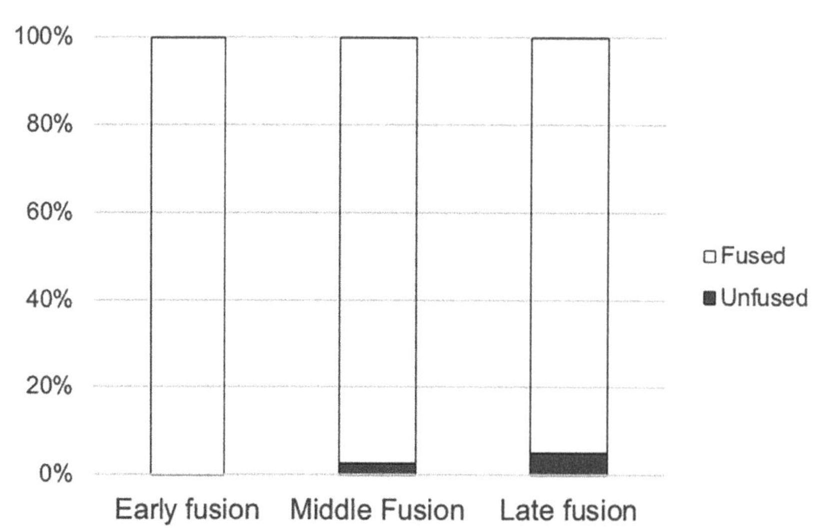

Figure 6.9 Epiphyseal fusion pattern in deer from Xinzhai

Table 6.10 NISPs of major animal taxa from layer and feature contexts in each phase at Xinzhai

Phase	Context	Pig	Deer	Cattle	Sheep	Human
1	Layers	21	12	8	7	0
	Features	136	42	51	6	73
2	Layers	455	235	126	158	31
	Features	491	225	90	69	362
3	Layers	233	124	81	55	0
Total	Layer	709	371	215	220	31
	Feature	627	267	141	75	434

between contexts (**Table 6.10**). As seen in **Figure 6.10**, where fauna from different phases were pooled together, while human depositions demonstrate obvious differences across these two types of contexts, the trends of NISP fluctuation in animals resemble. This resemblance in non-human animals was further statistically tested and the result of contingency analysis displays a high possibility of association between the two (χ^2 = 47.279, p (no assoc)=3.03E-10). To eliminate the illusion brought by lumping effect, NISPs in each phase, except for Phase 3 (no features were recovered during this phase) were tested respectively. The two trends in Phase 1 are somehow obscure, probably due to the small sample size. Statistically, however, they are associated (χ^2 = 15.537, p (no assoc)=0.001). In the same figure, the line of layer NISPs and that of feature NISPs in Phase 2 are almost identical, and a contingency analysis also suggests a close association (χ^2 = 37.288, p (no assoc)=3.9993E).

The results above suggest a similar accumulation of animal bones between layers and other feature contexts. It remains uncertain whether this statistical resemblance alone suffices to illustrate depositional agency and accumulation process. Lumping effect could have resulted in an inadequate reflection of differentiation in formation process.

Internal and external depositions

Seen from the spatial layout of the settlement, fortified walls served as an important landmark. The distribution of features inside and outside formed a sharp contrast in number and density. In the following section, I examine whether the formation process of depositions relates to its location in the settlement.

Using the inner wall serving as a border, depositions inside and outside can be compared. Comparing feature contexts, 97 feature contexts of these located within the inner city, 31 were from Phase 1, 66 from Phase 2 and none from Phase 3. 11 feature contexts were located outside the inner wall, two from Phase 1, nine to Phase 2 and none to Phase 3.

In Phase 1, the total NISPs of four animal taxa (pig, deer, cattle, sheep/goat), together with humans and the average NISPs in each feature deposit were calculated and plotted in **Table 6.11** and **Figure 6.11**. While contexts inside the inner wall contained far more bone specimens, none of the animal bones except pig bones were collected from features beyond the area. Seen in **Figure 6.11B**, pig is the only taxon found in features outside the wall to have yielded relatively higher average NISPs in external features than internal ones. While no human remains were found among external feature depositions, all human bones were deposited within this internal area.

Table 6.12 and **Figure 6.12** compare the faunal NISPs between inner and outer features from Phase 2. In Phase 2, the contrast in the number of bones between the inside and the outside remained dramatic. The average NISPs of animals in inner features varied among different taxa,while

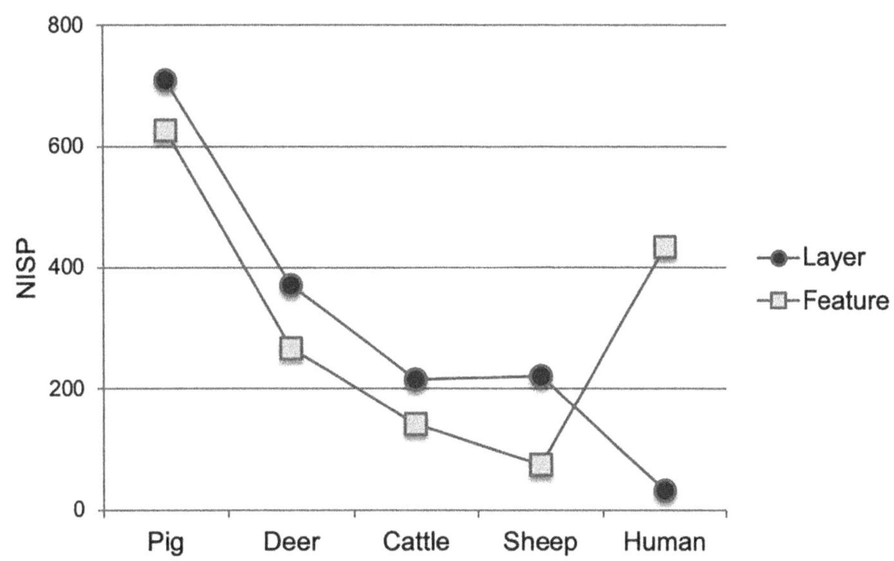

Figure 6.10 Comparisons of human and animal NISPs in layer and feature contexts at Xinzhai

Table 6.11 Comparisons of total NISPs and average NISPs per context between internal and external feature depositions from Phase 1 at Xinzhai

	Inside (31 contexts)		Outside (2 contexts)	
	NISP	Avrg. NISP per context	NISP	Avrg. NISP per context
Pig	127	4.1	9	4.5
Deer	42	1.4	0	0
Cattle	51	1.6	0	0
Sheep	6	0.2	0	0
Human	73	2.4	0	0

Table 6.12 Comparisons of total NISPs and average NISPs per context between internal and external feature depositions from Phase 2 at Xinzhai

	Inside (66 contexts)		Outside (9 contexts)	
	NISP	Avrg. NISP per context	NISP	Avrg. NISP per context
Pig	493	7.5	8	0.9
Deer	218	3.3	7	0.8
Cattle	83	1.3	6	0.7
Sheep	65	1.0	4	0.4
Human	422	6.4	13	1.4

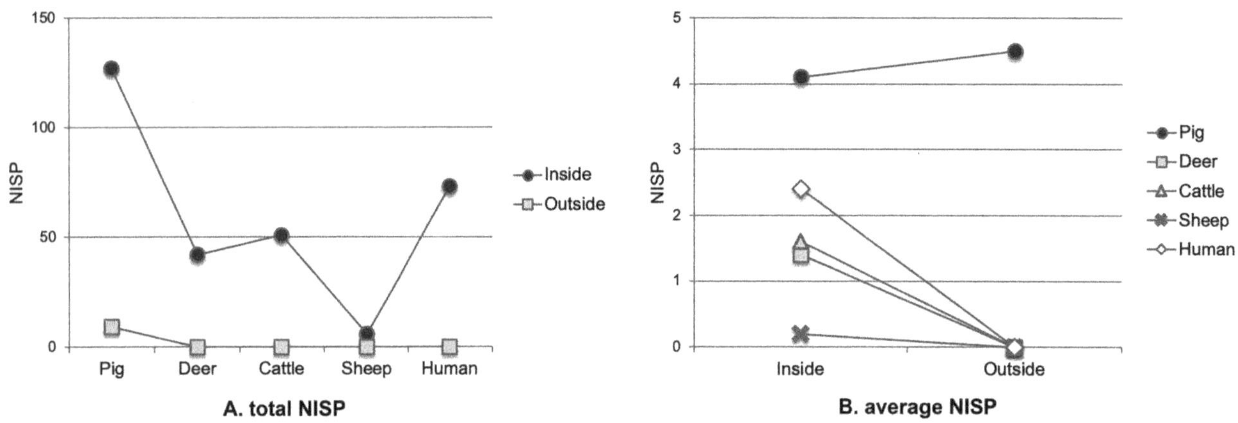

Figure 6.11 Graphic comparisons of (A) total NISPs and (B) average NISPs per context between internal and external feature contexts from Phase 1 at Xinzhai

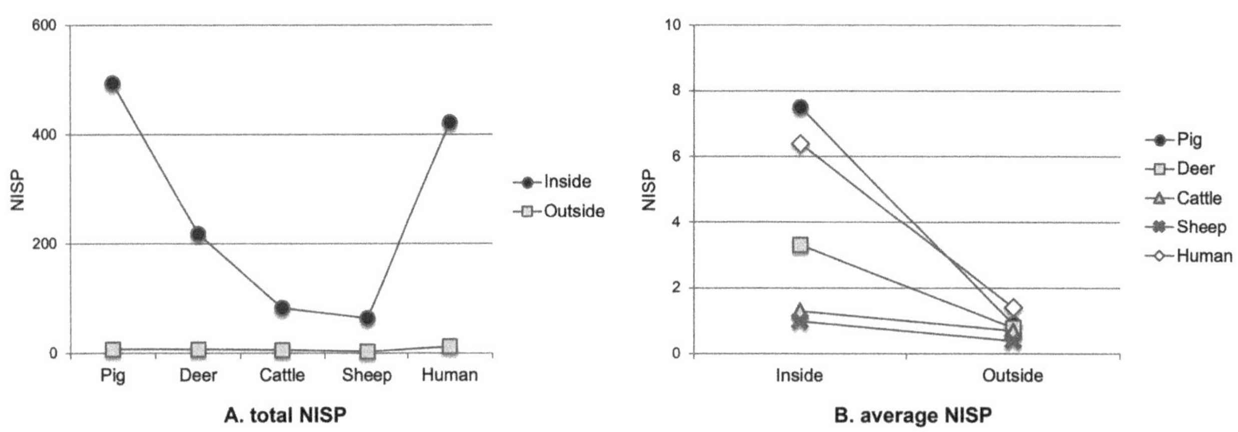

Figure 6.12 Graphic comparisons of (A) total NISPs and (B) average NISPs per context between internal and external feature contexts from Phase 2 at Xinzhai

those outside remained fairly similar. The majority of human remains were deposited in internal features, a distribution pattern largely corresponding to those of other animals.

In terms of depositions in layers, 15 contexts were located within the inner wall, of which six were from Phase 1, nine from Phase 2 and none from Phase 3. A total of 67contexts were located outside the wall, of which four belonged to Phase 1, 44 to Phase 2 and 19 to Phase 3.

Table 6.13 summarises contexts from Phase 1. Compared with the number of bones found in feature depositions, fewer bones were collected from layer contexts inside the wall. On the contrary, faunal remains deposited in layers outside outnumber the external features. We should

Table 6.13 Comparisons of total NISPs and average NISPs per context between internal and external layer depositions from Phase 1 at Xinzhai

	Inside (6 contexts)		Outside (4 contexts)	
	NISP	Avrg. NISP per context	NISP	Avrg. NISP per context
Pig	22	3.1	2	0.5
Deer	8	1.3	4	1.0
Cattle	6	1.0	2	0.5
Sheep	2	0.3	5	1.3
Human	0	0	0	0

Table 6.14 Comparisons of total NISPs and average NISPs per context between internal and external layer depositions from Phase 2 at Xinzhai

	Inside (9 contexts)		Outside (44 contexts)	
	NISP	Avrg. NISP per context	NISP	Avrg. NISP per context
Pig	43	4.8	412	9.4
Deer	27	3.0	208	4.7
Cattle	7	0.8	119	2.7
Sheep	8	0.9	150	3.4
Human	19	2.1	12	0.3

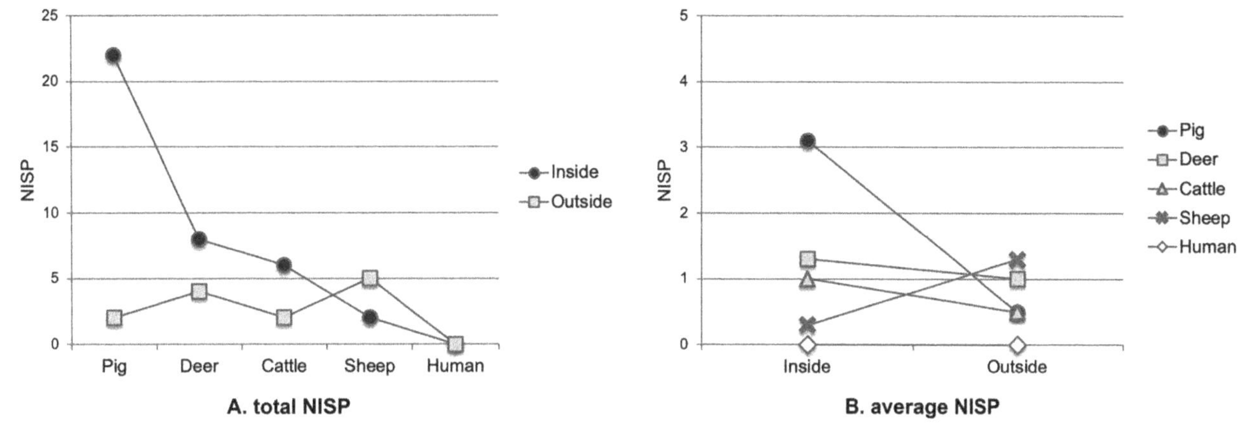

Figure 6.13 Graphic comparisons of (A) total NISPs and (B) average NISPs per context between internal and external layer contexts from Phase 1 at Xinzhai

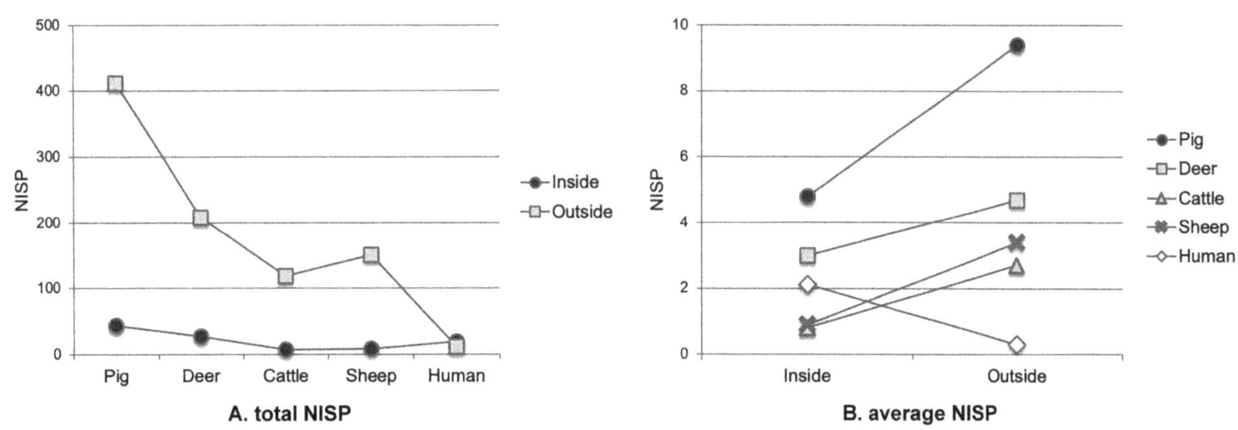

Figure 6.14 Graphic comparisons of (A) total NISPs and (B) average NISPs per context between internal and external layer contexts from Phase 2 at Xinzhai

note that there were slightly more sheep deposited in layers outside than those inside, unlike other animals. Correspondingly, layer context outside has higher average NISP of sheep (**Figure 6.13**). No human remains were deposited in layer contexts during Phase 2.

During Phase 2, NISPs in external layer contexts greatly outnumber those on the inside (**Table 6.14**). A Wilcoxon test also supports the differentiation between the two in terms of non-human animal depositions (p=0.07). Except for human remains, more bones were always in deposited in external layer contexts as opposed to internal layer contexts. For human remains, NISPs inside and outside largely remain the same. The average NISPs show a similar pattern in non-human animal assemblages, while on the contrary, among humans the average NISP was higher inside (**Figure 6.14**).

Table 6.15 Spatial distribution of pig-rich and deer-rich contexts at Xinzhai

Feature type	Percentage (%)	Number of depositions	
		Inner enclosure	Outer enclosure
Phase 1 features	Deer NISP ≥ 50	5	0
	Pig NISP ≥ 50	17	2
Phase 1 layers	Deer NISP ≥ 50	0	0
	Pig NISP ≥ 50	3	0
Phase 2 features	Deer NISP ≥ 50	6	2
	Pig NISP ≥ 50	38	3
Phase 2 layers	Deer NISP ≥ 50	1	9
	Pig NISP ≥ 50	3	10

During Phase 3, no layer contexts were recovered inside the wall, while 19 layer contexts were located outside. Among these were 233 pieces of pig bone, 124 deer bone, 81 cattle bone and 55 sheep bone.

Inferred from the numbers above, for feature depositions (*e.g.* pit), there seemed more animal bones deposited within the enclosed area, while for layer contexts, many more bones were buried outside. A sharp spatial contrast emerges.

In order to examine the correlation between spatial layout and depositions of pig and deer bones in particular, depositions were further investigated by comparing pig and deer percentages (**Table 6.15**). The number of contexts containing 50 per cent and higher deer or pig remains was counted. In Phase 1, most contexts (including layers) with 50 per cent and higher deer remains were located in the inner city. In Phase 2, a difference between layer and feature context appeared: in feature contexts, most pig depositions were located inside the wall, and large proportion of pig deposition in layers were from features outside. In the case of deer, in Phase 1 deer-rich depositions were situated within the walled area regardless of context type. In Phase 2, while internal features contained more deer-rich depositions, the pattern reversed in layer context, where deer-riches depositions were recovered from layers outside.

Young and adult depositions

In animals, such a distinct separation of the young and the adult was not obvious. It seems that depositions of animal bones were occasionnaly still referenced by age. Approximately speaking, those contexts deposited with juvenile animals were clustered in squares such as 2000T1 to T4, the area crowded with pits of various sizes and shapes. While 14 contexts filled with young animals that were located within the walled area, only six depositions of young animals were recovered outside the inner wall. Additionally, all were discovered from archaeological layers outside rather than from feature contexts.

Human remains were also examined to examine the existence of age-related differentiation. **Table 6.16** presents the demographic summary in different types of depositions.

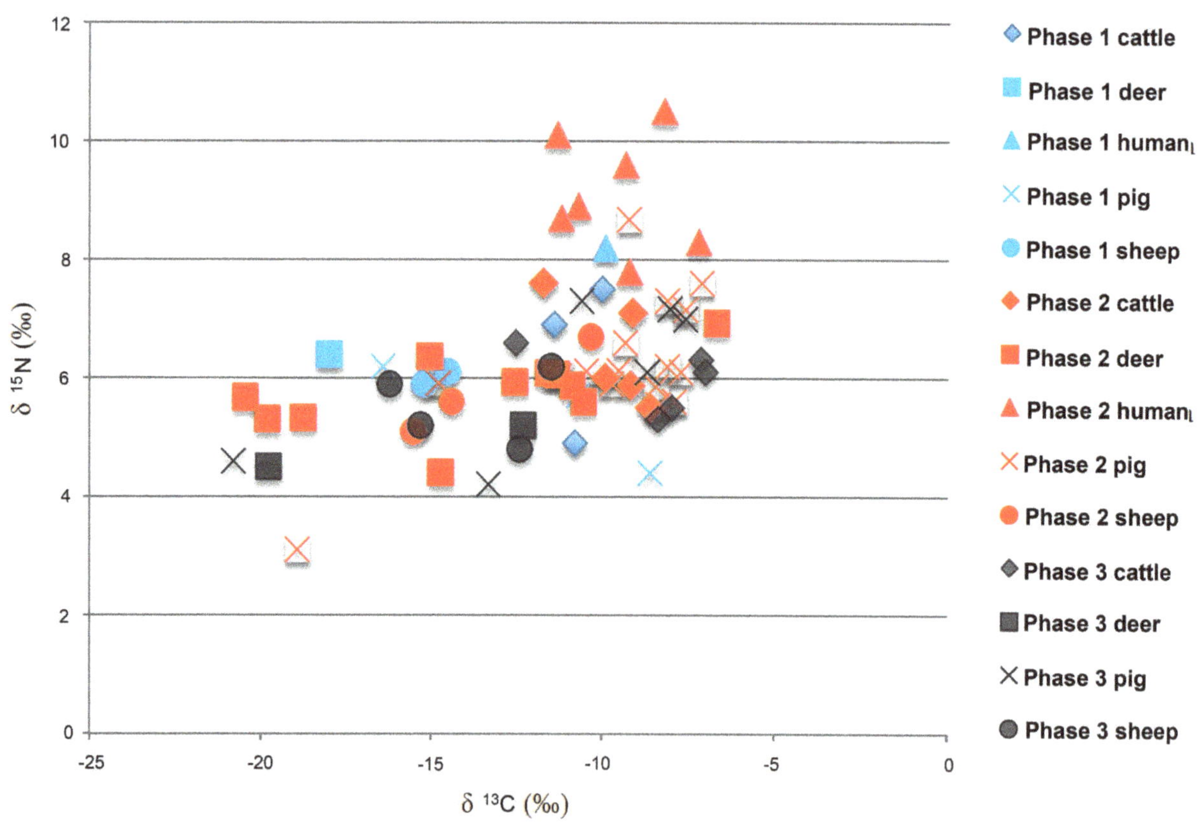

Figure 6.15 C & N isotope data of human and animals at Xinzhai

Table 6.16 Age distribution of human remains and their depositional types (numbers of depositions are compared)

	Adult	Child	Mix	Child versus adult ratio
Formal tomb	10	2	0	0.20
Layer	10	4	2	0.40
Pit	29	11	1	0.38

Two major depositional types are compared: formal tombs and other depositions including pits, ditches and layers. Numbers of depositions with different group of human remains are compared. As seen in **Table 6.16**, both adults and children were found in various types of contexts. The difference in number of contexts does not suggest a clear pattern. The ratio of child versus adult in tomb context is 0.2 while the ratio in other contexts has doubled to 0.4, indicating that children were more frequently deposited in contexts other than tombs. However the evidence does not yet support an obvious age division among human beings.

Type of deposition

The results of the above comparisons seem to suggest two contrasting sets concerning spatial division at a site-wide scale

Table 6.17 provides a summary of these depositions associated with their spatial differentiation. The inner wall marked the borderline separating those depositions that were located within the inner enclosure and those on the settlement outskirts. The difference in the number of features found in these two areas marks a sharp contrast. In terms of skeletal data, more bones (including human bones) were deposited in layers than in features in general. Human bones were collected largely within the walled area, the case being similar for young animals. Associated bone groups were found restricted to the inner enclosure. Animal remains largely accumulated within the wall. However, there appeared two contrast patterns of bone accumulations between feature and layer depositions. Though NISPs vary among taxa, the general tendency demonstrates an obvious spatial contrast among all animal taxa.

Table 6.17 Summary of characteristics of internal and external depositions at Xinzhai

		Inside the wall	Outside the wall
Number of features		97	11
Number of features deposited with human bones		44	7
Number of ABGs		2	0
Number of features deposited with young animals		14	6
Number of animal remains	In features	More	Fewer
	In layers	Fewer	More

6.3.6 Summary of isotope data

Due to the unavailability of bone materials, no new isotope data could be achieved for this project. Published isotope data from the same site is collated and summarised in **Table 6.18**.

Table 6.18 Summary of published isotope data from the Xinzhai site (data collated from ACSACPU & ZMIACR 2008; Dai et al. 2015; Zhang et al. 2015)

Phase	Taxa	Number of specimens	δ^{13}C‰		δ^{15}N‰	
			Range	Average	Range	Average
Phase 1	Cattle	3	-11.4~-10.0	-10.7	4.9~7.5	6.4
	Deer	1	-18		6.4	
	Dog	2	-11.1~-7.9	-9.5	5.9~7.3	6.6
	Human	1	-9.9		8.2	
	Pig	3	-16.4~-8.6	-11.9	4.4~6.2	5.4
	Sheep	2	-15.2~-14.5	-14.85	5.9~6.1	6
Phase 2	Cattle	5	-11.7~-8.6	-9.7	5.5~7.6	6.4
	Deer	11	-20.4~-6.7	-14.2	4.4~6.9	5.8
	Dog	1	-12.8		8.4	
	Human	7	-11.3~-7.2	-9.6	7.8~10.5	9.1
	Pig	16	-18.9~-7.1	-9.8	3.1~8.7	6.2
	Sheep	3	-15.5~-10.3	-13.4	5.1~6.7	5.8
Phase 3	Cattle	4	-12.5~-7	-8.6	5.3~6.6	6.0
	Deer	3	-19.8~-12.3	-16.0	4.5~5.2	4.9
	Dog	1	-9.7		5.8	
	Human	0				
	Pig	6	-20.8~-7.6	-11.5	4.2~7.3	6.1
	Sheep	4	-16.2~-11.5	-13.9	4.8~6.2	5.5

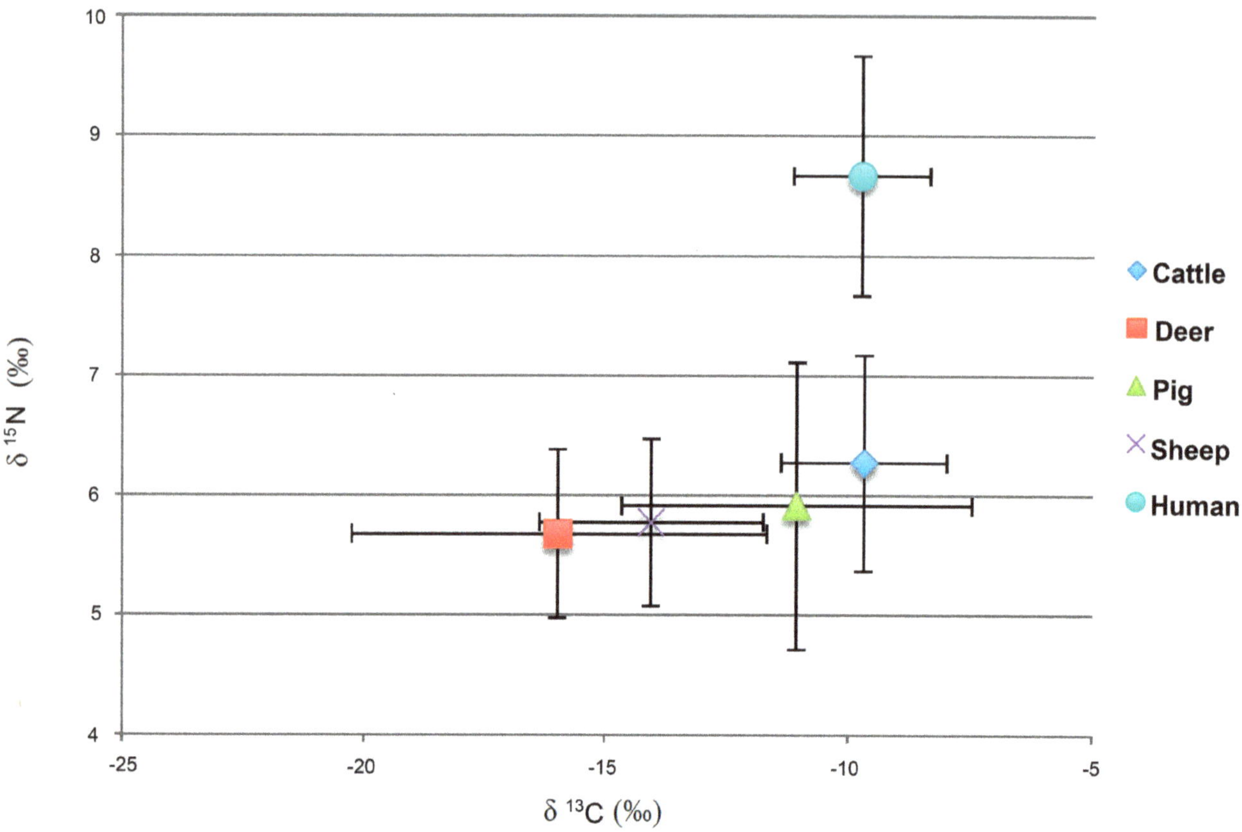

Figure 6.16 C & N isotopic range of human and animals at Xinzhai

Seen from **Figure 6.15 & 6.16**, the human $\delta^{15}N$ range is enriched from that of animals by approximately one trophic level, suggesting that human dietary intake might have contained animal products. Their $\delta^{13}C$ values ranging from -11.2 to -8.0 per mille, indicating consumption of C4 plants — very likely millet or millet byproducts — at the Xinzhai site. No differences can be detected among different age and gender groups. Cattle $\delta^{13}C$ values fall in a relatively narrow range and display strong C4 signals. The $\delta^{13}C$ values of pigs cover the range from -18.92 to -7.1 per mille, implicating a wide dietary spectrum. This wide range is not only the result of a broad dietary spectrum in individuals but also influenced by the individual diversity, seen from **Figure 6.16**, in which values displaying both C3 and C4 signals can be seen. The nitrogen isotope level in pigs is slightly higher than those of other herbivores — except for cattle — perhaps owing to their scavenging of human food residues. Sheep carbon isotope values, ranging from -16.5 to -11.5 per mille are suggestive of consumption based on a mixture of C3 and C4 plants. As individual variations seem moderate, the wide range was likely the result of shifting grazing over seasonal pastures with occasional supplementary fodder under special situations. Deer isotope pattern has the broadest range which displays both strong signals of C3 and C4 plants, and overlaps partially with that of pigs.

Isotope values of pigs and deer from different types of contexts are compared in **Table 6.19** & **Figure 6.17**. No apparent differences are found between pit and house depositions in general. If any, deer deposited in houses yielded slightly higher N isotope values, though the difference is subtle. The majority of pigs deposited within houses have isotope values indicating C4 plant consumption, but there is one outlier, which demonstrates lower values in both nitrogen and carbon isotopes. Due

Table 6.19 Summary of pig and deer isotopic data from various contexts (data collated from ACSACPU & ZMIACR 2008; Dai *et al.* 2015; Zhang *et al.* 2015)

Taxa	Context	Number of specimens	$\delta^{13}C$‰		$\delta^{15}N$‰	
			Range	Average	Range	Average
Pig	House	7	-18.9~-7.6	-9.9	3.1~8.7	6.1
	Pit	13	-16.4~-7.6	-10.4	4.2~7.3	6.0
Deer	House	8	-19.8~-10.5	-14.4	5.3~6.9	6.0
	Pit	6	-20.4~-6.7	-14.5	4.4~6.4	5.3

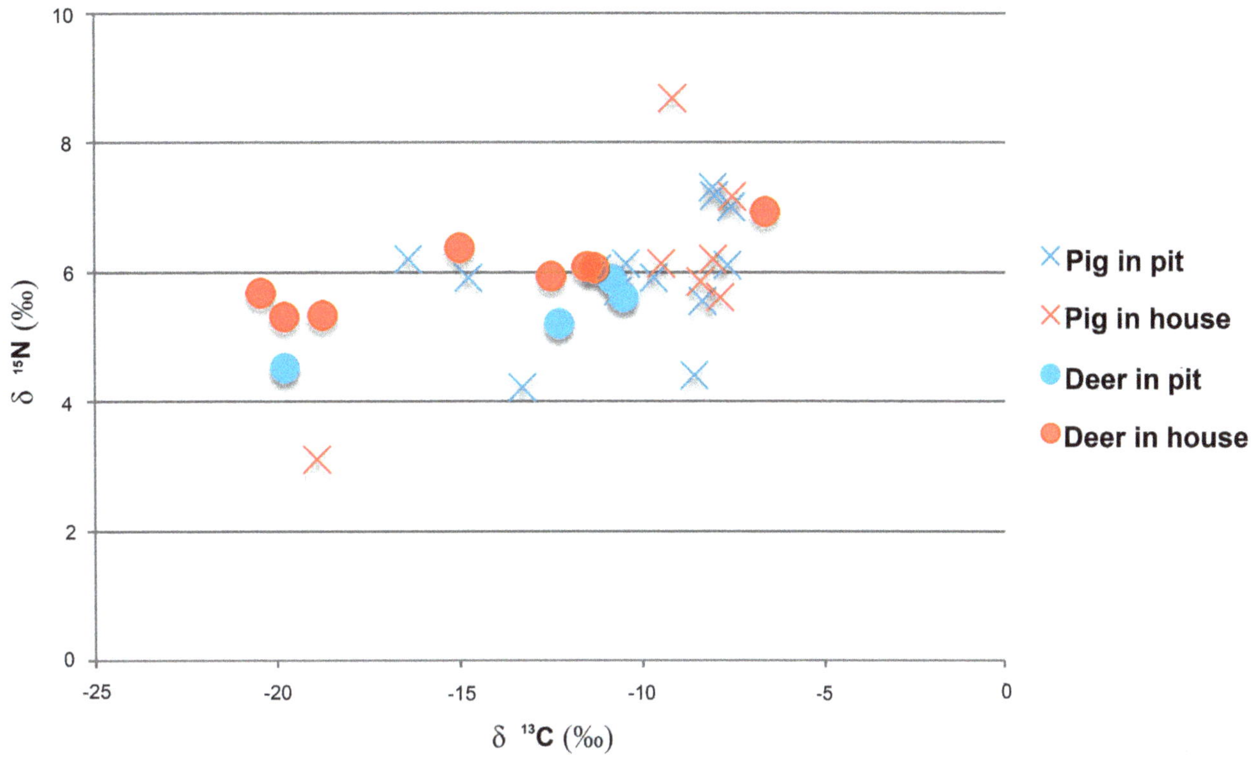

Figure 6.17 Comparing isotope values of animals from various contexts

to the small sample size, no concrete conclusion can be asserted so far.

6.4 Discussion

Inferring from the results above, this section begins with the investigation on the exploitation of animal resources with a detailed analysis of the interaction between deer and human beings. The status of deer then leads to a discussion of animal categories with regard to a conceptual map involving landscape, economy, politics and many more social domains of life at Xinzhai.

6.4.1 Exploitation of animal resources

Considering the relatively large collection of faunal remains, animal products seemed to play an important part at Xinzhai. This diversity of animal products from different taxa might have been utilised by residents at Xinzhai for various purposes.

Pig must have been one of the most important animals exploited. Pork was on the menu at Xinzhai. The majority of pigs were slaughtered while still sub-adults. Killing at the age between one and three years old assured a substantial amount was available for consumption as well as good quality of meat. Pigs were consumed in a large quantity. Although in latter phases supplemental protein from other animal sources kept increasing, the proportion of pigs remained stable through all phases. Such large herds of pigs were very likely free-range. Inferring from the broad distribution in isotope values, pigs might have had access to both wild plants and cultivated millet, and their options for food varied among individuals. They could have scavenged around the settlement for human food refuse. They could have also freely wandered into the fields adjacent to the village, rooting for truffles or the like. Cucchi and his colleagues (2016) combine geometric morphometric and isotopic approaches to investigate pig husbandry in ancient China. Their results suggest that pigs at Xinzhai were probably largely free-range pigs managed in extensive herds. They further underline the differentiation between these free-range pigs at Xinzhai and smaller pigs at Xiawanggang, another late Neolithic site in Henan Province, which were herded by households. The authors therefore speculate that elite-controlled husbandry were practiced at a large scale in Xinzhai (*ibid.*). In contrast to the pig management at the household level, this strategy was better for societies where a central authority exerted power the organisation of economic and political activities, such as an elite site like Xinzhai.

Alternatively, the diversity in isotope values suggests pig mobility. The Xinzhai site operated as the pivotal settlement within the regional network. It would not be surprising to anticipate the existence of the centripetal flow of resources and people towards Xinzhai. As a consequence, the occurrence of demographic migration and exchange of resources including animals might have happened at the site. Such exchanges would certainly have given rise to the presence of non-indigenous introduced animals. It seems unclear whether animals, if non-indigenous, were brought in to the site when while still alive or in the form of animal products. In the case of pigs, driving a herd of pigs over miles may not have been an easy job.

Cattle and sheep were domesticated animals introduced into China during the late Longshan (*ca.* 2,500 ~ 2,000 BC). Exploitation seemed to increase with time. At Xinzhai, both provided important animal products. Cattle, more vulnerable under severe conditions, might have been fed completely with millet fodder provided by shepherds. This extra care paid to cattle was proposed to reflect their ritual significance (Dai *et al.* 2015). Faunal evidence involved in this project, however, neither support nor negate cattle's distinct role in ritual events or other activities. In terms of sheep exploitation, Dai and colleagues (2015) postulate wool utilisation at Xinzhai by examining the age profile of sheep. This postulation can hardly be tested by the current project data in this project due to the lack of information pertaining on the age at death for sheep. Re-examination of the isotope evidence demonstrates a relatively broad range of isotopic values slightly biased to C3 plant intake in sheep. Sheep management might have involved rotational grazing in different patches of pastureland during different seasons and occasional provisions of fodder under particular conditions such as winter, birth season and so forth. The area between the inner wall and the intermediate moat might have been used as the pen to shelter herds, considering that a relatively large collection of sheep and cattle assemblages was recovered there. Seemingly, cattle and sheep were not penned in the vicinity of households. This pattern is roughly consistent with that found for pigs, loosely controlled presumably beyond the household scale.

In addition to these domesticated species, deer, conventionally perceived as wild animals, were also heavily exploited and gradually emerged as a major animal resource for for Xinzhai residents. The status of deer is discussed in the following section.

6.4.2 Status of deer

Deer, in general, are considered wild animal resources for exploitation. Given the growing body of bioarchaeological evidence and profound understanding towards a complicated human-animal relationship in the past, this viewpoint may need a drastic revision.

Manipulation of deer

Management of deer means that people intervene in the life of deer for many purposes and to various extents. Thus the life of a deer may not always follow a natural circle and changes may occur to their behaviour or further physical traits correspondingly as the result of human manipulation. Ranging from feeding royal swans on River Thames, control of stray cats, to raising herds, numerous examples in daily life suggest various ways of animal management.

Augmented by the isotopic analysis of Xinzhai deer, some scholars (*e.g.* Zhang & Zhao 2015; Dai *et al.* 2015) envisage a scenario where deer were manipulated by humans. Seen in **Figure 6.15** and **Figure 6.16**, there is no doubt that deer at Xinzhai did have access to C4 plants—most likely millet byproducts. The scale of consumption, however, might have varied and the variation in scale would have in turn determined isotopic signals in deer bones. The distribution of deer carbon isotope covers a broad range, indicating dietary variations among individuals. That is to say, according to their consumption patterns, three groups of deer can be roughly identified: those that foraged on wild leaves and grass, those with occasional access to millet products and those largely feeding on crops. As these deer belonged to different groups, their relationship with human beings might have been different as well. On one hand, deer could have been wild in nature, but raided the fields and fed themselves on millet byproducts opportunistically. Humans might have permitted their access to crop fields so as to increase the frequency of the encounters, as a way to amplify chances of prey procurement. Alternatively those deer that raided the fields might have been killed for the sake of crop protection. These ways of procurement are alike to 'garden hunting' recorded in ethnographic literature. It seems under both circumstances, crop consumption would have been occasional. If so, it remains doubtful whether the dietary intake would have been in commensurate scale to shifting isotope values in deer bones. In other words, deer bones demonstrating strong C4 signals must have had more frequent C4 plant intake, which occasional field-raiding would hardly have provided. There are chances that deer were either husbanded just the same as those domestic animals, or herds were deliberately free ranged within a relatively large patch of land for most time, albeit with a consistent supply of fodder. At times when they were required for specific purposes, deer would have been corralled and slaughtered.

It is plausible to envisage different methods being employed to procure deer from different groups. Generally speaking, deer were unlikely to be kept as 'domesticated' livestock, even those with instant access to crops. Inferring from the age profile, the youngest deer so far recovered at the site are indicated by a mandible from an individual under 17 months. The absence of infant or juvenile animals in the assemblage seems to argue against the hypothesis of domestication *sensu stricto* of deer involving human manipulation of breeding.

Garden hunting provides one hypothesis to explain the deer group with moderate C4 isotope signals. However, no evidence so far can securely point out the existence of garden hunting at Xinzhai. Garden hunting essentially is hard to identify by archaeological assemblages. It remains poorly understood in terms of types and quantities of animals caught in agricultural fields in various regions under different ecological and cultural conditions. A large body of literature on garden hunting dwells on ethnographic observations of pre-industrial societies in America or tropical rainforests (*e.g.* Linares 1976; Greenberg 1992; Neusius 2008), further limiting the application of those conclusions to other archaeological cases. Therefore a series of questions must be considered with caution in the case of Xinzhai.

In the first place, the number of wild animals recovered at Xinzhai is extremely small (NISP=15) once we exclude the deer assemblage. Although deer such as sika deer, muntjac deer and Pierre David's deer are well-known crop raiders, there is also a long list of other animals including wild boar, rat and fox which pose threats to farmlands as well (Conover 2001: 107-11). These wild species, however, were poorly represented at Xinzhai, especially compared with the relatively abundant deer remains. This argument somehow weakens the 'garden hunting' hypothesis. Nevertheless, one may speculate that small wild mammals might not have been on the residents' menu at Xinzhai whereas venison was preferred gourmet. This preference for food finally led to the preservation of deer bones in the site. Especially in Xinzhai – a farming society distinguishable from those horticultural societies where garden hunting serves as an alternative to animal domestication – there might have been many food choices, and consequently small game were not in demand. They might have been disposed wherever killed, very likely in fields outside the settlement. However, size seemed not to be the primary criterion to select animals for food. Though other small mammals were probably not consumed, there were also a certain number of bones belonging to fairly small sized deer, as seen in the deer assemblage.

Another line of evidence worthy of our attention concerns the population of over three years of age that was targeted. There are chances that among such prey, the three- to four-years old group was included. In wild populations, this group comprise mainly sub-adults which have already left their mothers but are not that experienced in leading an independent life. They are more vulnerable and as a result fewer efforts are required to hunt them down (Clutton-Brock *et al.* 1982: 178). This targeted group basically reflects expedient hunting strategies rather than opportunistic encounters with game. Though lacking systematic information on the age structure of prey of garden hunting, research on the mobility of sika deer in modern Japan does not imply any preference to farmlands correlated to deer age and gender (Takatsuki *et al.* 2000). The possibility that the age profile dominated by sub-adults and adults mirrored the structure in the living group can be precluded. It seems reasonable to speculate that deer sneaking into agricultural fields might vary in age. If protecting cultivated crops served as the primary aim of Xinzhai residents, people should have caught all deer that were a potential threat to crops regardless of age, and in consequence diversity in the age profile would be anticipated. The current age evidence, however, contradicts this speculation. In addition, the selective cull of young animals, in particular young male individuals, would have boosted the production of local populations according to the modern observation (Fletcher 2000: 137), which appeared in conflict with the original aim to protect crops from damage caused by wildlife.

Thirdly, location of agricultural fields might have mattered. In their recent research on the wildlife damage to farmlands in natural reserve areas in west central China, Xu and colleagues (2016) point out that wild animals tend to raid agricultural fields located adjacent to forests and distant from settlements, reducing the risk of being caught, as exemplified by the Wolong National Natural Reserve in Sichuan Province. Palaeoenvironmental study at Xinzhai suggests the existence of broad leaf forest and moor landscapes near the settlement (ACSACPU & ZMIACR 2008: 509). How far agricultural fields were situated away from the forest remains unclear. Fields might have been situated outside of the enclosed settlement in the vicinity so as to maintain agricultural fields with easy access and under close guard. Such a postulation currently lacks archaeological support.

The evidence above appears pitted against 'garden hunting' on a large scale. In other words, garden hunting was unlikely the main strategy for procuring deer. A hypothesis of management to varying degrees seems more plausible. Deer were sort of free-range as indicated in their isotopic values overlapping with those of livestock such as sheep and pigs. The diversity amongst individuals by no means would implicate fenced control and foddering. Bringing deer herds under management can simply be achieved through taming the dominating leader of the herd, a similar way practiced by modern reindeer herders (see Ingold 1980). Occasional provision of fodder and shelter from severe weather or particular situations (*e.g.* giving birth), and constant provision of fodder might have taken place. As herds were not penned within the enclosure, the absence of young individuals would be reasonable. Instead deer herds might have spent most of their time wandering in forest and fields adjacent to the settlement. Some seventeenth century records show that it was fairly easy for a forest keeper to call together free-ranged deer in England (Fienne 1888, cited by Fletcher 2000: 137). People selectively culled adults and sub-adults deer groups at Xinzhai, as seen from the age profile. The patterning of body elements unlikely supports the transportation of particular body parts. Deer might have been brought back to the site either on hoof or as a full carcass. Managed deer could have served various aims and been procured in diverse ways.

Purposes of manipulation

All the evidence above implies management of deer for a certain purpose. Then what were these deer managed for exactly?

First and foremost, human beings have a long history of venison consumption. There was no exception for this region, where deer were exploited throughout history. Unlike other sites in the same region, Xinzhai singles out evidence pointing to the potential manipulation of deer herds. What would be the motivation to drive this change in deer exploitation? Some scholars adopt an ecological perspective, interpreting it as a response to the environmental deterioration of that period (ACSACPU & ZMIACR 2008). This argument did not hold up under the light of isotope data. Venison could have supplemented meat resources, securing an adequate supply of protein. Nonetheless there seemed no point in sharing crops with deer to ensure meat on hoof instead of simply hunting completely wild populations. In modern days, when venison has become a popular dish, deer are farmed for industrial meat production. According to the figures provided by the New Zealand meat industry, farmed deer are slaughtered at 12, 18 and 27 months to ensure meat quality (FAO 1982). Xinzhai deer were butchered at slightly older age compared with these modern farmed deer. This would appear to be an acceptable delay, taking both the quantity and quality of meat for different demands between the past and modern societies into consideration.

Beyond their function as food, were there any other roles deer might have played? The proposal of royal deer hunting (Dai *et al.* 2015) provides an alternative insight into deer management and exploitation at Xinzhai. Royal hunting became a salient recreation and ritual activity of the later Shang dynasty, as evidenced by the considerable number of oracle inscriptions making mention of the practice (Fiskesjo 2001). By making associations with such royal hunting, Dai and colleagues (2015) propose that deer management at Xinzhai serviced elite hunting as well. This postulation, though inspiring and exciting, requires supporting evidence.

A look into body part representation and tool assemblage at the site may provide some clues. At Xinzhai, the distribution of body element in deer differs from pig, while in resemblances can be found between deer and sheep or cattle in certain cases. Even if generally agreed that pigs were slaughtered on site, it is unnecessary to assume that deer were butchered outside the settlement and transported back. Other factors including storage andconsumption pattern would have altered the distribution. Therefore I would avoid any hard conclusion from the body part representation, whose patterns are insufficient to support either logistical transport of particular body parts or consumption of whole carcasses.

Age profiles suggest the exploitation of sub-adult and adult individuals around three years old. If hunting as a means of deer procurement had been practiced at Xinzhai, animals in this group would have been the primary targets to exhibit the strength and skill of the hunter while securing sufficient quantity of meat. This thread of evidence only explains the hunting strategy and says nothing about management of deer for hunting. The target game, being unnecessarily managed deer, could have been entirely wild as well.

Tool assemblage yields rich information about how animals were procured. Arrowheads occupy an intermediate proportion in the tool assemblage, and their percentage increased over time. On one hand, arrowheads were presumably weapons used for hunting. On the other hand, they could also kill large mature animals regardless of status. An analogy can be sought in guns. Guns are ideal weapons for hunting nowadays and a convenient method for farmers to dispose of a large domestic animal. Therefore the abundance of arrowheads does not essentially equal with the demand for hunting.

Although evidence suggests that deer at Xinzhai might have been managed, it remains unclear whether these deer were managed specifically for the hunting purposes recorded in several ancient texts. *Vice versa*, although hunting seemed to be the most feasible method to procure deer at Xinzhai, prey deer were not necessarily managed. Despite the obscurity surrounding this deliberate management for hunting, it is apparent that the management of deer later took a different trajectory from other domestic animals such as the 六畜 *liu chu*, the six typical livestock taxa recorded in pre-Qin texts such as 爾雅 *Erya (Ready Rectifier)* and 左傳 *Zuo Zhuan (Commentary of Zuo)* which refers to pig, horse, cattle, sheep, dog and chicken. Though deer were never recorded as livestock in ancient texts, archaeological evidence implies that they were managed to various degrees. That deer were excluded from livestock did not necessarily mean that they were included in this wild category. The wild/domestic dichotomy may turn out to be much more porous than conventionally perceived. In the case of Xinzhai deer, it seemed that neither of the two categories fits this status. Deer category somehow wandered in limbo, another strong argument for the porosity of the wild/domestic dichotomy. 'Deer hunting' is probably one example elucidating the hazy boundary.

Hunting and deer category

Although evidence to illuminate any direct link between deer management and hunting remains insufficient, hunting as a method of procurement, might have contributed to the categorisation of deer. In some cases of animal classification, instead of animals themselves, it is the treatment of the animal that matters in terms of its categorisation. Linguistic examples support this argument. The word-stems preserved in many Indo-European languages today are considered as proxies for lifeways in the ancient world (Gimbutas 1952, 1970). For example, many terms in Indo-European languages denote domestic animals, while the names of plants and fish are almost absent. This sharp contrast encoded in our lexicons is believed also to reflect the heavy exploitation of livestock such as pig, cattle and horses on the European steppe (Lehmann 1962: 208). In addition to the direct lexical reflection of reliance on domestic animals, other aspects of the lifeway on the steppe were imprinted into lexical stems indicating the processing methods of food rather than the food itself, taking plants for example (Jones 2002). The stem '*grno-*' referring to 'wearing down a particle' is the root of grain, grind, corn and kernel; '*bhreu-*' meaning 'heating, boiling and fermenting' is used in words such as bread, broth, brew, fry and ferment (Buck 1949 & Partridge 2002, cited by Jones 2002). The importance of plants on the proto-historic European Steppe was recorded in word stems that signal the treatment of plants, rather than plants themselves. In many ways, the treatment of the animal and plant organisms themselves played the crucial role.

As we learn from textual records, hunting was the most frequent way to obtain a deer. In his study of 186 pieces of Shang oracle inscriptions describing hunting events, Guo (1982: 198-9) found that deer were the most frequent and abundant prey type, mentioned as many as 24 times. Among the 24 records, one showed as many as 384 individuals were hunted down in one event (*ibid.*). It appears plausible that we can speculate a close association between deer and hunting.

Other than hunting, there might have been many ways to kill an animal. Nevertheless, the way to kill deer was always referred to as hunting. If the method of killing is indicative of animal category, it is worth exploring the meaning behind 'hunting'. There are some linguistic hints. *Tian 田*, referring to 'hunting or chase' (Schuessler 2007: 496) was a frequent word in the Shang oracle bone inscriptions related to royal hunting (Fiskesjo 2001). *Tian 田*, pictographically depicting an organised-cultivated-field, also meant agricultural fields. Fiskesjo (2001) is convinced that the use of *tian 田* implies a 'significant relation between agriculture and royal hunts, which may involve the forced conversion of the wild into farmland'. The interplay of the two meanings encompassed in one word precisely demonstrates the porous boundary between wild and domestic categories. Unlike the biological division, wild and domestic categories might have been mutually interchangeable in ancient China. Via hunting, the wild characteristic in an animal was transformed into the subordinating domestic virtue. Hunting seemed to serve as the metaphor of alteration as illustrated by the many hunting events arranged for Shang kings to demonstrate their power to conquer, as recorded in oracle inscriptions. Further examples can be found in later dynastic periods. As mentioned in *太平御览 Taiping Yulan (Imperial Reader)* of the Song dynasty, animals were managed in some royal parks for the sake of elite hunting and display. In these cases the hunting option as principally a performance, a drama, in which the wild characteristics of animals were purposefully constructed and amplified through the practice of hunting, and where their domesticated side — manipulated by humans to a certain extent — was mitigated. This pre-arranged management and procurement of animals in the royal hunting grounds again reinforced the interchangeability between wild and domestic categories.

Returning to Xinzhai, there seems to be have existed more than one group of deer. The grouping was marked by dietary differentiation, which was not, however, reflected in archaeological depositions. There might have been various methods of exploitation corresponding to these different groups. Hunting could have been one of those methods. As details are scant on how hunting events were arranged in the past, no conclusion can be firmly asserted about the relationship between hunting and the deer category. It is at least clear that deer were the primary prey targeted by hunting. The tradition of deer hunting rather than deer rearing continued and was emphasised throughout history despite the management of deer in certain cases. This tentatively implies that deer might have some connections to both 'hunting' and 'domestication'. The connection to Xinzhai may seem faint, but is in turn a convincing explanation of the hazy boundary between the wild and the domestic.

6.4.3 Age as a classificatory filter

Age divider seemed to be non-significant in the classification of both human and animals, although traces are still found that demonstrate the difference in treatment of different age groups.

In terms of human, while there are recoveries of proper tombs for children, the number is much lower than for adults, as presented in **Table 6.16**. This table tentatively supports an age divider in mortuary aspects. On the other hand, in non-mortuary contexts, there were also skeletal fragments belonging to adult individuals were scattered in pits which also contained other animal bones, ceramic and lithic fragments. However, even in these, most likely, trash conditions, child skeletal fragments seldom ended up at depositions alongside adult human bones. The scarcity of mixed deposition, to some degrees, implies an age-related separation in dealing with rubbish. It will be recalled that there is a pervasive binary classification concerning the valuable versus valueless in classifying objects in general. In other case studies dealt with in this project, further classifications of animals seemed to only have been practiced among those that had been grouped under the valuable category. At Xinzhai, an age divider, rather than being employed to classified animals for ritual or ceremonies, was effective in dealing with valueless rubbish, manifested in the separation of children and adult bones in refuse depositions.

Similar examples are found in the rubbish disposal of the Endo of Kenya. As recorded by Moore (1982; 1986: 102), the Endo follow a set of complex 'rules' to discard ash, chaff and animal dung in particular locations corresponding age and gender connotations. Endo discard is not however practiced with full awareness of symbolic meanings (*ibid.*). Thomas (2012) comments on Endo rubbish disposal and underlines that this activity, though carrying symbolic implications, is by nature habitual, unevaluated and without deliberation. That is to say, people simply operate in this way as it is 'appropriate' in that culture. The symbolic meaning is not particularly performative. A similar scenario might have taken places at Xinzhai. Members of the Xinzhai community disposed of children and adults separately. They did so as a result of a certain social institution. The separation itself was never aimed at purposefully strengthening difference. If applicable to Xinzhai, then it should be noted that in this particular case of Xinzhai, age as a classifier was not reinforced by the patterning of rubbish disposal, unlike the cases of Wadian and Wangchenggang, where age categories were deliberately on display. Rather, the behavior at Xinzhai was but a simple social institution without purposeful construction of categories.

There seems no sign of an age filter for classifying non-human animals. Though some depositions with more young than olders were found within the inner city, differentiation between internal and external depositions was far from convincing. Most refuse depositions were filled with bones from both young and adult animals, without deliberate selection and organisation.

6.4.4 Classification of animals and spaces

No one would deny that there a variety of classifications are involved in our life. The same can be said of life at Xinzhai. Though the focus of this project remains on the classification of animals, it cannot be ignored that such a classification was rooted in a set of broad and profound principles through which many more objects and activities were involved. These principles, were firstly established on the basis of the proximity of objects to the member of the society.

Spatial layout and area themes

Manifested in the spatial layout of the settlement, 'proximity' to society members was embodied by the establishment of walls and moats demarcating the enclosed residential area from surrounding unoccupied fields and the unexploited distant wilds, thus formulating the contrast between the 'inside' and the 'outside'.

The outmost moat together with the Shuangji River formed the settlement outline. The inner wall delineated the central town area. The town appeared to be the principal area for conducting economic and social activities. Few archaeological features and remains were recovered in the area between the outer and inner walls, including a handful of pits along with a number of faunal remains (largely sheep and cattle). This peripheral area, not intensively occupied, seemed to be have been utilised as a zone for certain agricultural related activities. More specifically, domestic herds were probably penned here overnight or under severe seasonal conditions, inferred from the relatively greater number of sheep and cattle bones. That the Shuangji River coursed through the area back in ancient time also provided a vital instant water source, especially for cattle.

Within the town area, though the separation of spaces was not that rigorous, the primary theme of each area can be recognised. The central north part containing its large open-air structure was presumably an area related to ceremonial events. No matter whether the structure was exactly referred to the ritual performance area known as a 坑 *keng* or 壇 *tan* in later dynastical texts, such a massive structure without a roof was unlikely suitable for everyday dwelling. Though risky to designate the feature as a specific ritual ceremony recorded in ancient passages, given the thousand year chronological gap in between, archaeologists have indeed recognised traces of burning and intentional burial of complete pig skeletons at the structure, potentially indicating a function relating to special events. On the other hand, the nature of the ritual activities occurring here, if any, remains debatable, in that the structure itself, though distinguished from other houses, does not boast a significant departure from other contexts in terms of age, gender, body element distribution of its fauna. The same can be said about pig and deer discovered at this structure, of which the isotope values turn out to be similar among individuals excavated from house and pit contexts. This result appears to echo a proposal of doctrinal ritual, a type of placid ritual practiced at a regular frequency (Reinhart 2015). With doctrinal ritual embedded into quotidian life, it would not be unexpected that animals involved in such practices were those same ones subject to daily utilisation.

Approximately 50m to the south of this large structure, the area (where squares 1999T1, 1999T2, 1999T5, 2000T9, 2000T10, 2000T13 & T12 were located) was clustered with houses, pits, heaths and tombs, particularly in Phase 2. This area was largely immersed in the household sphere. Houses were likely constructed for daily dwelling, as suggested by the presence of postholes and hearths. Burials were placed adjacent to or underneath houses. Although it is hard to articulate each of the tombs with particular houses, they might have somehow been connected in both spatial layout and symbolic connotation corresponding to kinship. Fragmented human remains from adults and children were seldom found outside of formal tomb contexts in this area. Individuals, in this sense, were incorporated into the household even when they passed away.

Further on in the southern part of the excavation area, pits were clustered in crowds. They overlapped with each other, seemingly lacking organisation and arrangement. Fragmental bones from children were largely recovered in the south margin of this area whereas a small number of adult skeletal fragments were deposited randomly in the area. It seems likely that adult human bones were located closer to the household area, while child remains were deliberately disposed further away to the north. In a similar vein, a greater young animal individuals, represented by juvenile pigs, were found concentrated in the south. Given this patterning of faunal remains, age categories seemed also to be expressed spatially, reflecting principles of inclusion and exclusion, though the spatial division of age was not explicitly demarcated, probably due to the porous nature of animal categories.

Categorical separation and conceptual segregation

The spatial evidence so far demonstrates that the layout of Xinzhai was guided by and in turn affirmed the principles concerning a broad distinction between 'inclusion' and 'exclusion'. This is seen from the construction of walls and ditches to separate the enclosure from the nature, and the proximity between houses and tombs to privilege the centrality of the membership within the community. Lexically, 家 (*jia*, referring to house, family and to domesticate) and 野 (*ye*, referring to field, wildness) broadly resonate the spatial division at Xinzhai. These oppositional structures, however, are not strictly impermeable. Instead under certain circumstances where the space was rather ordered in gradations towards the core of the cultured society, penetrating a neatly-cut binary separation. A typical example comes from the three-layered design of the fortified enclosure. Though walls were indeed built up to symbolise the distinction between the 'internal' and the 'external', the contrast was constructed through the graded spatial arrangement in which the rural area between the outmost moat and the inner wall might have served as a buffer area between the external natural world and the internal domestic settlement. Moreover it was very likely that areas outside the moat were used as agricultural fields, again blurring the division between *jia* 家 (domestic) and *ye* 野 (wild). The layout of the inner town with the area probably pertaining to ritual activities further added to the complexity. The main theme was still sketched under the opposition while the contrast sets, comprising three units (ritual, household and rubbish areas), appeared to have been structured in a spatial pattern in grade.

Integrating the faunal evidence, the opposition between *jia* 家 and *ye* 野 was seen moving beyond spatial arrangement and into animal categories. Exemplified by the case of deer, deer were managed and at the same time hunted. Deer might have been placed in both wild and domestic categories. These apparently conflicting ways of arranging the deer category again brings us back to the debate pertaining to the status of deer. The tension between *jia* 家 (domestic) and *ye* 野 (wild) seems sufficient as an interpretation for the changing status of deer. Managed deer may have been cared for, but the deer had never been a member of the domesticates. On the contrary, deer managed for hunting, if any, had to be wild conceptually, because only 'wild' animals could be hunted. One does not need to 'hunt' but instead 'kill' a domestic animal. As a result, the care paid to deer had to be deliberately neglected to emphasise their nature being wild and kill a deer through hunting acted as the symbol of the domestication of wildness. At this point, it is clear that the oppositional contrast between *jia* 家 and *ye* 野, the domesticated and the wild, was interchangeable through an appropriate way of alteration, in this case, which is hunting. These two concepts were interdependent. *Ye* 野, the wild side, was constructed in relation to the domesticated *jia* 家, ultimately serving to strengthen and privilege the central place of the 'domestication'. Hunting down a wild deer and bringing it back to the settlement epitomised control of the domestic over nature, the wild others. In this sense, the wild performed merely as a catalyst pushing domestication to its culmination. In a similar vein, walls were erected to separate the cultural sphere from the natural sphere. The separation, however, was centripetal, as values were only put on one side and the construction of the separation aimed to witness the alteration (the process of domesticating) of the wildness through both physical and metaphorical means.

Categorisations of animals and spaces merged together under the tension between *jia* 家 and *ye* 野. These two contrast principles can be further dissected. *Ye* 野 has a triple meaning. In the first sense, it denotes field and the physical space. The second entry denotes the 'wild, particularly in the ecological sense. This ecological definition further was entitled with political implication and developed into the connotation of 'savage' with particular reference to people beyond control. In later dynastic periods, the loose control denoted by *ye* 野 was frequently used against the central authority *chao* 朝 (which with deep roots in the context *jia* 家 to be discussed shortly). For instance, 在野 (*zai ye*, literately in the field) was used to describe people and parties out of power. Slightly differing from *man* 蛮, *yi* 夷 and *hu* 胡, the concepts latterly develop to depict barbarians distinct from civilised people under the control of the empire, when the principle of *ye* 野 is used, is the potential for controlling rather than the conflict in between that is emphasised and implied. From this point of view, deer at Xinzhai fitted nearly the category of *ye* 野. The lexical concept of *jia* 家 is also packed with multiple meanings. In the sense of house and building, it is contrasted with *ye* 野 from a spatial perspective. In terms of ecological implication, *jia* 家 is often used as an adjective to describe the status of being domesticated. For instance 家猪 *jia zhu*' refers to the domesticated pig. In the last sense, *jia* 家 can also be understood as a verb meaning 'to control'. At this point, *jia* 家 and *ye* 野, the former referring to control, the latter being controlled, are again articulated.

The *jia/ye* 家/野 dichotomy was reflected by the spatial layout and animal categorisation at Xinzhai. This separation was gradually taking shape and further developed, subsequently forging a series of key concepts that have turned out to profoundly influence the historic course of Chinese civilisation. For instance, the idea of civilised *zhongyuan* 中原 (the middle kingdom or the central state) was always defined and revised against its corollary *yi* 夷 on the periphery (Fiskesjo 1999).

Last but not least, the concepts of 'domesticated' and 'wild' in the modern biological sense must be cast aside in order to understand the categories of *jia* 家 and *ye* 野 in the context of Xinzhai. The definitions of 'domesticated' and 'wild' categories are steeped in modern biological discourse, of which the utilities are restricted within sorting plants and animals. At the Xinzhai site, *jia* 家 and *ye* 野 categories being far more versatile, were applicable to a wide range of tangible objects, empirical experiences and abstract ideas in many domains. In fact, these categories formed a pivotal part of the perceptual map where everything sensed, experienced and constructed were organised according to people's understanding of the world. This order of the organisation, in turn, directed the interplay between various factors within the site, thus determining the future trajectory of society.

6.5 Summary of animal categories at Xinzhai

The colossal scale, multiple layers of ditches and walls, large open-air structure, and chronology of transitional period between prehistoric and dynastic periods are the features of the Xinzhai site. No wonder it has become a hot topic among archaeologists keen to resolve questions on the emergence of Chinese civilisation. Putting aside all these debates, this chapter has made the very first attempt to fit the animal category into this complicated picture.

Livestock were heavily depended upon. Pigs, with little surprise, still played the main role. The husbandry strategy employed, however, suggests a shift towards free-range management involving the kind of large-scale authority-based control proposed by Cucchi and colleagues (2016). The same could be said about sheep and cattle, which were also maintained in large herds, probably beyond the household level. It seems that an authority was sufficiently powerful to take every aspect of life under its control.

Age divisions were not as clear as for the previous two sites. These might have played a role in the classification, though the evidence remains weak. At Xinzhai, the age filter appeared to have only tentatively worked for humans rather than animals, and the age divider seemed somehow obscure.

One of the most interesting discoveries of the Xinzhai faunal assemblage is the possible existence of deer management. It has been long taken for granted that deer are wild animals, without further clarifying the definition of 'wild'. Isotopic and skeletal evidence at Xinzhai points to varying degrees of deer management and the possibility of hunting in some cases. This scenario best explains the categorisation of deer. In terms of management, deer were no longer wild animals. Meanwhile people ceased to intensify management of deer, resulting in different trajectory for this creature as opposed to other livestock. As indicated by textual information, hunting remained the major method to kill deer. Hunting served as a metaphor of alteration from the 'wild' to the 'domesticated'. Inferring from this example it may be tentatively speculated that Xinzhai residents might have already been aware of both wild and domestic categories, the boundaries between which, however, were porous. These contrasting sets did not only reside within animal categories but were also applied to many social domains such as spatial construction, agricultural production, and so on. In my later analysis, I extend this division onto a much broader conceptual segregation that was pervasively manifested in economy, politics, language, ideology and so forth. A detailed discussion follows in the next chapter.

7

Animal Categories: a Synthesis

As discussed in earlier chapters, animal classifications in pre- and proto- historic China were a component of a larger cosmological system that thoroughly permeated the institutions and thoughts of past societies. The late Neolithic constitutes a complicated stage in China on which various lines of politics, economy, social stratification become intertwined. These lines and knots keep interweaving and finally weave out the picture of the early Bronze Age, which in turn leaves a notable legacy to later societies. Animal categories were both shaped by and they themselves shaped the perceived world within this picture. In this chapter, observations from the three archaeological sites discussed above are compiled into a larger archaeological background. I attempt to situate the animal classification within the panorama of the social landscape from the Longshan to the Shang.

7.1 Human versus non-human classification

It is difficult to estimate the exact time when the category of 'human' appeared. Suffice it to say that the appearance of special treatment of the human body including construction of tombs embodies the distinction of a human category. It is beyond the scope of the current study to trace the origin of a human category which might date as early as the Palaeolithic. Narrowing down the focus to the research period, we note with interest that not all biological humans (*i.e. Homo sapiens*) were included in the indigenous category of 'human'.

To begin with, it is essential that we define the meaning of 'human'. Rather than in the biological sense, the definition of the human category should be sought through ancient eyes. One direct way to define the human archaeologically is through mortuary practice. The deceased were interred in tombs and, if necessary, funeral events were held for them. In most cases, animals would not be treated in this way outside of certain situations when they were perceived as similar or equal to human beings.

Seen from examples demonstrated in this study, children, very likely, were excluded from the category of 'human', as proper child tombs have only been discovered rarely. This might imply that they were excluded from the local category of 'human being' within the study region. Apparently biological traits are not the only criteria to define 'human'. In many societies, children can only become human after their rites of passage, so did the Neolithic and Bronze Age communities. In fact, we are still faced with the same question to define when we become 'humans' in the present day. Modern societies hold various opinions in the debate over the definition of human: Dose life begin at conception or birth? When do teenagers reach adulthood? All such questions matter because the answers to them determine many critical issues: legitimation of abortion, moral dilemmas in science, and conviction of juvenile delinquency, to just name a few.

In addition, multiple human skeletons were found in certain type of pits, exemplified by the 'sacrificial pit' at Wangchenggang (see ch 5). Some scholars believe that they were captives or slaves from conflicts (HPIACR & MCH 1992; Reinhart 2015), though evidence for violence is, frankly, inadequate. In ancient China, captives and slaves ranked the lowest in the community. Sometimes they were even excluded from members of the community, seen as beast-like semi-humans (Fiskesjo 2001). The same can be said of children. It is possible that the category of 'human' was in fact the category of community member. Members of the community were entitled to be 'human' whereas a non-member was not qualified to be a 'human'. The core to the category resided in defining 'membership' rather than 'humanity'. The exclusion of children, captives, slaves and the like continued after the Neolithic and was even reinforced in later periods. In the late Shang dynasty, captive *Qiang* people, an ethnic group formerly living in the northwestern marginal region of the Shang territory, were given animal nomenclature, signaling their removal from the human category and inferiority to Shang people (Fiskesjo 2001). Even nowadays, the term *quan zi* 犬子 (literally my dog-like son) is used to denote one's son to show the speaker's humility. Boundaries between animals and humans thus were not delineated by biology.

The category of 'human' was very likely established upon the recognition of membership within a community. Criterion for the inclusion within the category was rigid, as not only non-humans (animals) but also less-humans (children) were not qualified. This category privileging the membership of the community was consonant with the strengthening of the principle of *jia* 家, discussed in the following sections.

7.2 Age categories

Age categories were distinctive among the animal classifications in the Neolithic and early dynastic periods. Juvenile animals were often distinguished from their adult counterparts. The division was applied to human beings as well.

7.2.1 Age classification and chronological change

The age classifier had appeared by late Neolithic Longshan. As earlier sites are beyond this research period, it remains unclear whether the age classifier had an even earlier origin. Age-related classification seemed overwhelming, being located in all the three sites involved in this book, although the manner of representing age categories varied in each case.

Juvenile pigs were used in foundation pits at Wadian, while adult pigs were found scattering across refuse depositions. Age distinction was represented by the segregation of different age groups. In addition to spatial isolation, the distinction in context (*e.g.* ceremonial events, daily consumption) corresponding to each age category was also recognised and contextual differentiation in turn solidified these categories (see ch 4). In contrast, age-related categories at Wangchenggang were characterised by deliberately bonding the young and the adult together within the same deposit (see ch 5). There were both pig and deer double pits recovered, and in each case an adult animal was buried together with a juvenile individual. Two age categories—the juvenile and the adult—appeared in the same deposition. Such selection could only be made after the specific classifier had been established. Presenting the two age categories of one animal taxon together in a single deposition seemed to fuel a negotiation between difference and similarity. In one sense, there existed two categories determined by age. On the other hand, allowing the time for growth, these two categories were indeed permeable. Boundaries between age categories were therefore porous. At Xinzhai, age categories were strikingly apparent in the attitude towards human death: adult humans were formally buried in proper tombs, whereas children, most commonly, ended up in rubbish dumps, with occasional exceptions (see ch 6). Differing from the other two sites, age categories were not preserved in the patterning of non-human animal remains.

Chronologically examining evidence from these three sites, no particular temporal pattern was observed in age-related classifications. Speaking very vaguely, towards the end of the Longshan period, the age filter became slightly weaker in the categorisation of humans and animals exemplified at Xinzhai. The Shang likely witnessed a contrary scenario in oracle bone inscriptions, where age distinction had become a widely adopted and major criterion to select animal sacrifices (*cf.* Wang 2007). The late Shang dynasty depicted in oracle inscriptions and the research period of this study are not separated by too great a stretch of time. It remains unclear whether there was a revival reinforced age categories in non-human animals during the interim, or alternatively whether the invisibility of age-filtered-non-human animal groups is merely an illusion due to the absence of evidence. What is clear is that age had served as a long-lasting criterion to classify animals throughout the late Neolithic and the beginning of the Shang dynasty in central China.

7.2.2 Age classification and depositional context

Age categories were also pervasive and were observed in various contexts. Age categories were indicated in ritual context. Animals were separated by age in sacrificial contexts, exemplified by the juvenile sacrificial pigs deposited under the house foundation at Wadian and the double pits containing one juvenile and one adult animal at Wangchenggang. Additionally, age as a classifier also appeared in refuse contexts, guiding the disposal of animal and human bones that had lost value. This scenario, though hazy, was found at Xinzhai where young animals — likely discard from food consumption and craft production — were thrown relatively far away from household area, and where remains of children were rarely mixed with those of adults.

The appearance of age categories in different contexts suggests that age was a strong and ubiquitous classificatory filter. Though age was pervasively applied in various contexts, subtle but significant differences remained. In the case of ritual ceremonies, age categories functioned to underpin a significance distinct from other contexts. The ritual context, in its turn, strengthened division on the basis of age. Thus the reinforcement was mutually interactive. This reinforcement in ritual context was intentionally presented by the particular presentation of age-related pattern. This was different from the situation at rubbish contexts. The patterning here was not deliberately made visible but the result of habitual behaviour according to social institution.

We have seen that age categories occurred across two sharply contrasting scenarios. The first was the dedicated presentation of animals and their categories with reference to the age filter, along with the symbolic meanings of these creatures. The other scenario in the refuse context could be seen as a repetition of 'habitual and unevaluated acts' without particular elaboration of material representation (Thomas 2012). These two archaeological patterns reaffirmed that age was a powerful classifier applied to both valuable and valueless things.

7.2.3 Age classification, language and ancient texts

In addition to archaeological evidence, age-related classification is also manifested in language. Especially in the Chinese writing system, a binary classification modified by age is observed. The character *shi* 豕 refers to pig while the character *tun* 豚 means piglet. Other word pairings include: *ma* 馬 (horse) and *ju* 駒 (foal), *yang* 羊 (sheep) and *gao* 羔 (lamb), *niu* 牛 (cattle) and *du* 犢 (calf). Its significance is also supported by the nomenclatural principle. Age categories are labeled with primary lexemes — one character standing for one category. Secondary lexemes borrow words to comprise the categorical names exemplified by those categorical names so as to distinguish domestic animals from wild animals (*e.g. ye niu* 野牛 bison, literally 'wild cattle'; *ye ma* 野马 wild horse). Categories indicated by primary lexemes are terminal and monosemic. Therefore the use of a primary lexeme may suggest a stronger classifier with an earlier origin, compared with those expressions of secondary lexemes. Age, inferring from the lexical evidence, was most likely a strong classifier.

Figure 7.1 Pictographic illustration of *tun* 豚 (piglet)

Following this linguistic hint, in the case of pig in particular it was found that the character *tun* 豚 denoting 'piglet' implies a close association between age category and ritual ceremony. The character *tun* 豚 was interpreted by the scholar Xu Shen (許慎, ca. AD 58~147) of the Han dynasty in *Shuowen jiezi* 說文解字 *(Explaining the Graphs and Analysing Characters)*, as depicting a piece of meat for ritual offering held in a hand (**Figure 7.1**). Characters from other ancient texts also display similar pictographic forms (**Table 7.1**). The pictographic composition of the character probably displays a ritual event whereas the character itself denotes 'piglet'. In this way, young animals and ritual performances were connected, further indicating the application of age-related classification in ritual contexts. Supporting evidence is also found in ancient textual passages that mention age as a criterion for selecting proper animals as sacrifices. In the *Li ji* 禮記 (*Book of Rites*), the book recording ritual performances and social forms of the Zhou dynasty according to the understanding of the scholars from the Warring States period and the Han dynasty, the Zhou preference for using juvenile sacrifice animals in a series of ritual ceremonies is mentioned. In the *Ji fa* 祭法 (*Law of Sacrifices*) section of the in *Li ji*, there is a line of text describing the use of a juvenile horse to as an offering to the Earth:

燔柴於泰壇，祭天也；瘞埋於泰折，祭地也，用騂犢

With a blazing pile of wood on the Grand altar they sacrificed to Heaven; by burying (the victim) in the Grand mound, they sacrificed to the Earth. (In both cases) they used a red juvenile. (tr. Legge 1885)

Mentioned in the section of '*Wang zhi*' 王制 (Royal Regulations), only juvenile bulls are qualified in the ceremony to worship Heaven:

祭天地之牛，角繭栗

On the bulls used in sacrificing to Heaven and Earth, the horns were (not larger than) a cocoon or a chestnut. (tr. Legge 1885)

Zhou peoples might have believed that the innocence of young animals made them suitable fodder for sacrifice to Heaven (Cao 2008). One may remain skeptical about the symbolic implication assigned to young sacrifices, given that *Li ji* was in fact complied during the Warring States period and early Han dynasty (*ca.* fifth to second century BC), more or less reflecting the nostalgia of scholars of later periods. Consequently these texts are secondary references to the Zhou ritual cannon and can hardly be used as direct evidence. Any incautious exploration of the symbolic meaning accredited to certain age category may lead to some untenable interpretation of ancient ideology. If we shift our focus to the consequence of such categorisation, it transpires that there is plenty of room for further discussion.

7.2.4 Age classification: consequences and correlates

One direct outcome resulting from the application of age classification was the use of young animals as sacrifices. Considered pragmatically, offering young animals in ceremonial rites minimised the owners' economic loss, especially when ritual offering of young animals did not involve consumption of their meat. Young animals are vulnerable and require extra care and investment, but have not fully grown to provide animal products of satisfactory quantity and quality. For this reason, a cull of young animals ends up as a compromise, fulfilling the task of offering sacrifice while reducing the actual financial loss.

Table 7.1 Different styles of Chinese character *tun* 豚 (piglet)

Oracle characters selected from *Jiaguwen bian* (Sun 1965)	Bronze characters recorded in *Jinwen bian* (Rong 1985)	Seal characters recorded in *Shuowen jiezi* (Xu 1963)

This scenario is well illustrated by the case of Wadian. However, towards the end of the Longshan period, young animals had become less frequently used and a diversity of animals tended to be involved as sacrifices in ritual ceremonies, as seen in the double pits of both young and adult animals dating to the very late Longshan period and the Erligang period at Wangchenggang. Later, in the Shang dynasty, young animals were still in use as sacrifices, though their selection was additionally modified by sex, colour, size and feeding strategy (Allan 1991; Keightley 2000; Wang 2007). It seems that age categories in sacrificial systems had been shifted through time. This temporal pattern correlated with the changes in husbandry strategies.

Offering young animals as sacrifices out of an economic concern is compatible to the animal husbandry strategy practiced at the household level. In this mode of husbandry, the household directly invests in and receives a return from livestock in their possession. Benefits and losses are their own. The owner has to be cautious about balancing religious return and secular investment. Consequently it would not be unsurprising that young animals became ideal candidates as ritual offerings. Infering from dietary resemblance between animals and human beings through isotopic signals, as well as smaller tooth size (Chen *et al.* 2015; Cucchi *et al.* 2016), household-level husbandry might have been practiced at Wadian. The medium size of dwellings, along with their spatial distribution, seemingly indicates that economic and social activities were conducted at the household level. Taking all these evidence into account, the preference of young animals for ritual sacrifices, partially for the sake of cutting down the investment, broadly fitted into the household economy at Wadian during the Longshan period.

Gradually moving from the Longshan period to the Erlitou and Erligang, despite an age classifier continuing to be adopted in ritual ceremonies, emphasis on young animals faded. Available examples at hand include double pits on the house floor recovered at Wangchenggang, where one adult pig along and one piglet was inhumed within the same deposition, usually associated with house structures. Those adult animals were fully-developed individuals, buried intact and left with no traces of consumption. If economic value was the primary concern, it would not have been ideal to offer adult animals in their prime days. The economic concern was losing its weight in determining the proper animal type for sacrifice. An even more complicated set of criteria involving judgments based upon colour, sex, age and so on was established in the late Shang, and can be seen in a number of oracle bone inscriptions (Wang 2007). This change occurred alongside the economic shift in animal husbandry from household-scaled management towards increasingly centralised control of livestock. The Shang society is commonly conceived as a 'very hierarchical society with a strong central authority, specialisation and division of labour' (Shelach & Jaffe 2014). As the Shang elites had high demands for animals serving for a variety of aims, it would be otherwise impossible if there had not been specialised groups in charge of animal management (Yuan & Flad 2005; Okamura 2008; Campbell *et al.* 2011; Campbell 2015;). Under this mode of institutionalised husbandry, economic losses due to religious demand is shared by all members of the society and thus the burden laid upon one particular household could be mitigated. Only when animal husbandry has been practiced under the institutional regime could a glut of sacrificial candidates be guaranteed, as was the case of Shang ritual animals. The beginning of this centripetal course had its traces at the pre-dynastic site of Xinzhai, where animal husbandry along with other economic and social activities was likely beyond the household level, as we have inferred from the settlement layout and isotope results.

To sum up, since the late Longshan, age acted as a strong divider in categorising animals. It remained one of the salient criteria for animal classification throughout the Shang and Zhou, particularly in aspects relating to ritual use of animals. Not only does the patterning of archaeological fauna suggest the existence of age dividers in various contexts ranging from ceremony to rubbish dump. Age-related categories were also preserved in the Chinese lexical repertoire, which continues to offer insight even if it cannot serve as direct evidence. Age classification was sometimes associated with ceremonial rites. The preference for young animals as ritual offerings might have helped their owner minimise economic investment under household-scaled husbandry. When the husbandry gradually became practiced under institutional control, other criteria were adopted in addition to age when classifying animals for ritual purposes, as we see in the subsequent Shang and Zhou.

7.3 Domestic-versus-wild division

The domestication of animals and plants dates very early in Chinese prehistory (see Zhao 2011 for plants; Yuan & Dong 2019 for animals). There is substantial evidence demonstrating the confident practice of domestication. One can hardly doubt the existence of domestication as a physical practice. Nonetheless such practice should not be conflated with domestic category as a concept perceived by past societies. Whether or not a wild/domestic dichotomy existed in the past remains open to doubt.

The case of deer at Xinzhai is a typical challenge to the wild/domesticated dichotomy. Xinzhai deer were seemingly anchored to neither a domestic nor a wild category. The category of deer was consistently refined by human management and procurement, implying blurred boundaries between the wild and the domestic. Despite such porosity in classification, there were still two opposing principles constructing a general contrast broadly corresponding to the wild/domesticated division during the beginning of the Bronze Age. These contrasting sets were illuminated by archaeological evidence from Xinzhai. As seen in chapter eight, I use *jia* 家 and *ye* 野 to address this contrast and situate the Chinese classification on an emic perspective. The *jia/ye* opposition was treated as a conceptual guideline, of which the structure was not necessarily binary and static. As we shall see shortly, boundaries between the principles were fairly porous.

This particular approach to binary principles is inspired by Hodder's study on Neolithic Europe. In his book *The Domestication of Europe* (Hodder 1990b), Hodder envisages a material and mental domestic/wild opposition during the Neolithic Europe, when processes of domestication involved controlling not only animals and plants but also people, thus creating new economic, political and social networks (Sherratt 1991). To re-contextualise this opposition, he adopts Latin and Greek words with Proto Indo-European linguistic origins such as *domus* (in its sense of 'house'), *foris* (indicating spatially outside) and *agrios* (hence 'agriculture') to rephrase the domestic and wild sides (*ibid.*). Hodder's view on 'domestication' beyond its economic notion *sensu stricto* to cover certain ideological implications is inspiring. The possibilities that the contemporary Chinese language, though having been substantially amended and re-invented during the course of transmission and development over the last thousands of years, has its origin deep in time, hence cannot be easily dismissed. In this context it is time to turn to the terminology and look further into the etymological history of *jia* 家 and *ye* 野 correlating with archaeological evidence.

7.3.1 Etymological history of jia 家 *and* ye 野

In many dictionaries the word 'domesticate' is linked to Latin *domus*, Proto-Indo-European *dom-* and *dem-*, Sanskrit *damah*, Slavonic *domu*, all generally implying house and family. It is surprising to find that, in Chinese — a Sino-Tibetan language distinct from Indo European language family — the adjective *jia* 家 describing the status of domestic animals likewise refers to 'house' in one of its many dictionary entries. *Jia* 家 denotes 1) house, 2) home, 3) family and household, 4) 'my' when used politely for family member, 5) domesticated, 6) classifier for families, business and companies, 7) school of thought, 8) a person with certain occupation or social standing, 9) a specialist in certain activity or field and 10) a surname (" 家" 2016a). There are further meanings specific to China's historical background of China. *Jia* 家 can refer to the manor of a minster or an officer of high status, especially during the Eastern Zhou Dynasty. In addition, it refers to court as used in *Lüshi Chunqiu* 吕氏春秋 (*Spring and Autumn of Master Lü*) (*ibid.*). *Jia* 家 also means capital, explained by Zhen Xuan (鄭玄, AD 127~200) in his book annotating *Liji* 禮記 (*Book of Rites*) (*ibid.*). In *Hanshu* 漢書 (*Book of Han*), *jia* 家 directly denotes the emperor and is employed as an adjective to describe peoples serving the emperor (*ibid.*). Finally, *jia* 家 is used in certain circumstances to indicate royal control and domination. The fact that in the long Chinese history, dynastic changes were *de facto* reshufflings of families in power casts light on the extension of the meaning of *jia* 家 to imply broader social and political units such as capital and dynasty. The etymological core of *jia* 家 resides in 'home' and 'family', from which the connotations are forged to include capital and manor spatially and court and emperor politically, to just name a few extended meanings. The conceptual trajectory witnesses the growing expansion of the idea of *jia* 家 beyond the household to larger social groups. The use of *jia* 家 to describe the subordinate status of domestic animals also indicates the power of control extended from the coverage of human beings to non-human animals, being a salient part in the process of expansion. The principle of *jia* 家 consists of two key concepts. Firstly it marks the distinction between 'internal' and 'external'. Secondly it suggests the opposition between 'individual' and 'communal'. These two contrasts are vital to our understanding of the creation of animal categories under the *jia*-versus-*ye* scheme.

Ye 野, though translated into English as 'wild', possesses meanings beyond those covered by the English word. Resembling the meaning of 'wild' in English, *ye* 野 denotes a status opposed to 'domesticated' and refers to fields away from the settlement (Schuessler 2007: 561). These entries roughly correlate with *agrios* as indicating a wildness linked to 'agriculture,' and with *foris* in which it refers to 'outside' and renders the word 'foreign' in the Proto-Indo-European language family. Mandarin boasts three other meanings of *ye* 野. It also means 'not in power,' 'unofficial' as opposed to 'court' and 'authority' which are both derived from *jia* 家. At this point, a play involving the distribution of political power was made on the *jia/ye* 家/野 contrast.

It is interesting to note that *ye* 野 itself probably means 'boundary' originally (Schussler 2007: 561). We may speculate that the meaning might have been forged in opposition to *jia* 家. From this perspective, the *jia/ye* contrast must have been so pervasive as to be preserved in the language. *Ye* 野 was sometimes used to indicate rude and violent behaviour. The etymological history recorded in the *Shuowen Jiezi* 說文解字 illustrates that *ye* 野 initially refers to fields outside settlements and is used as an adjective to describe rude and uncultivated people. Modern Mandarin dictionaries contain more entries under the character, among which the meaning of 'wild' as contrary to 'domesticated' to describe animal status seems a particularly late attachment. This again discerns the differentiation between the wild/domestic categories in the biological sense and the *jia/ye* 家/野 contrast in the past. I adopt the terminology of *jia* 家 and *ye* 野 in order to avoid projecting our modern pre-existing judgments upon ancient communities and thus to address the issue from the optimum emic perspective.

7.3.2 Expansion and intensification of jia 家

The etymological trajectories of these two characters clearly feature an expansion of the idea of *jia* 家 and its influence upon the making of *ye* 野. The expansion and reification of ideas pertaining to *jia* 家 is supported by archaeological data.

The house was the very centre of the *jia* 家 principle during the middle Neolithic central China. At sites such as Tanghu in Peiligang culture or Jiangzhai in Yangshao culture, household economy might have played the central role in social production (Li 2010; Peterson & Shelach 2012). The formulation of the notion of *jia* 家 through the household gradually extended outwards so as to bring the whole settlement under its domain. The most obvious trait was the fortification of settlements, highlighting the perceptual segregation by spatially demarcating the 'insider' and the 'outsider'. There was an abundance of fortified settlements appearing from the late Longshan period. In central China, the late Neolithic sites such as Guchengzhai, Wangchenggang and Xinzhai were enclosed by walls and/or moats.We have studied the second and third of these sites in studied in chapter five and six. The idea of *jia* 家 was initially hammered down at the household scale. The connotation of *jia* 家 was gradually expanded to embrace broader social units and diverse economic productions. The expansion was reflected not only in spatial enlargement but also in delineation of boundaries between communities. Walls and moats took responsibility for demarcating 'inside' from the 'outside'. Simultaneously, within the settlement, the architecture layout with the palace complex separated from other functional areas, marking discrimination among social groups within. In this way, the order and ranking within the *jia* 家 sphere was created and clarified in the form of materialised representation, reifying the intangible concept of *jia* 家.

During the Shang dynasty, the principle of *jia* 家 was not only concerned with the real world, but also the abode of the dead. Ancestral worship reached its zenith at that time and an imagined community of the ancestors became 'the most immediate unit of social and political identity' (Campbell 2009). All these bundled relationships of living and the dead justified the legitimation of *jia* 家 beyond the secular world.

Beyond the settlement scale, the reinforcement of *jia* 家 was manifest in the reaction to a more extensive and complex inter-regional social network, involving migration, trade, intellectual transmission, violent conflict and so on. These traits might have been rooted in the earlier Longshan period, but were more clearly illustrated in the Erlitou period. Liu (2004: 235) recognises the inter-regional Erlitou network, focused upon the 'production and distribution of prestige goods' including bronzes. She underlines the interaction between core and peripheral areas; the former was charged the craft production while the latter was responsible for provision of raw material (*ibid.*). This core-periphery system can be re-interpreted under the *jia/ye* 家 / 野 opposition. The subordinate peripheral area was on the side of *ye* 野 which provided raw material — unworked natural resources. Meanwhile the dominating core area transformed the raw materials into dedicated items — bronze vessels, for example — by the technology and labour under the control of *jia* 家.

This process of *jia* 家 expansion was apparently continued and intensified under the Shang. With the ongoing process, it became important to confirm the membership of the community and to stress the powerfulness of the community. Ancestral cult thus was a prominent way to privilege one's identity in the community. In terms of animal husbandry, a strategic shift from household management to institutional control suggests that the economic aspect of the *jia* 家 principle was upgraded to a more extensive and intensive scale. Therefore the process of this expansion was both centrifugal (as the expansion was outwards) and centripetal (as power agglomerated inwards).

Now let us return to animal categorisation. Wadian wild boar were deposited along with morphologically domestic pigs without intentional segregation. Similarly, at Wangchenggang, no traces of treatment reflecting wild and domestic differentiation in faunal remains could be found, though walls were constructed indicating a spatial differentiation between inside and outside. It was not until Xinzhai that the *jia/ye* 家/野 contrast became straightforward. At the same time, animal categorisation started to reflect some wild/domestic differentiation. Exemplified by the exploitation of deer, managed but possibly killed by hunting. Deer displayed characteristics from both domestic and wild sides. Hunting served as the method to swift between the two. The deer category could not be fixed statically to either of the two. That is to say, boundaries between the wild and the domestic were fairly porous. Perhaps it would be better to understand the category of deer on the spectrum of scale with one end on the wild side and the other the domestic. The allocation of the category on this spectrum was highly mobile depending on the context of utilisation and people's intention to emphasise either trait. The contrast between the wild and the domestic manifested in animal categories reached its peak during the late Shang dynasty, when royal hunting became one of the prominent symbols to mark the conquest of domestic power over the wild.

7.3.3 Structure of jia/ye 家 / 野 *opposition*

The *jia* 家-versus-*ye* 野 opposition was not constructed equally on both sides. For example, construction of living space was emphasised on areas within the settlement. The domestic sphere formed the core of *jia* 家. It was also true within the regional network, where merit accrued around the central settlements ranking at the top of the hierarchy.

In terms of animal classification, effort was made to elaborate the wild side. In Xinzhai example, deer were very likely managed. Although direct evidence is lacking letting us clarify whether deer were particularly managed to satisfy the demand for hunting, ancient textual records seem to support this hypothesis. Hunting was a critical procedure for reclaiming the wild characteristics of the animal. It was also through hunting that the wild nature was transformed into the subordinate tameness under the power of *jia* 家. Threads can also be found lexically. *Tian*

田 means agricultural field. Pictographically the character depicts cultivated fields organised in a square shape. It also refers to 'hunting or chase' (Schuessler 2007: 496) relating to royal hunting inscribed in Shang oracle bones (Fiskesjo 2001). The two facets of the word imply the intimate association between the wild and the domestic which seemingly involved a conversion between the two (*ibid.*). Rephrasing the conversion in my terminology, that is the transformation of the *ye* 野 to *jia* 家 principles. In contrast, though there might have been some changes in husbandry strategy of livestock, no evidence has been seen so far that was devoted to elaborating the construction of the domestic category *per se.*

This tendency continued through the Shang dynasty. The proportion of wild animals increased dramatically in the Yinxu period in the latter half of the Shang. At the great settlement of Anyang, a considerable number of wild animals including bear, bison, bird, tiger and so on were recovered, especially within the palace-temple area (de Chardin & Young 1936; Young & Liu 1949). Some of these animals live deep in the forest and far away from human settlements. These are not tame creatures. The royal hunts, as suggested in oracle bone inscriptions, were indeed the means of bringing these animals back into the settlement. In Fiskesjo's (2001) article on royal hunting of the late Shang dynasty, he concludes that the hunt operated as a crucial tool to show off the king's 'domesticating potency' (Campbell 2015) and legitimise his power. This was one of the means to add prestige to the power of the *jia* 家 principle held by establishing the category of wildness on its opposite side.

Wild categories were also constructed on bronzes. Animal motifs are frequently seen in bronze vessels, which feature beasts including tiger, elephant, rhinoceros, ox, bear, bore, sheep and deer, aside from bird and turtle (Chang 1981). These depictions of wild animals granted people general but still vague impressions about the wild categories. Meanwhile familiarity was restricted and its distance from quotidian life was deliberately maintained by engraving wild animals on vessels for specific events rather than daily utilisation in order to retain the 'wildness' in contrast to the domestic sphere.

The domestic side was constructed in a less intensive way. The augmentation of *jia/ye* 家/野 differentiation led to the establishment of rigid rules on selection of animal sacrifices for ceremonial rites in the Shang dynasty. Fiskesjo (2001) found that in most situations concerning ancestral sacrifices and mortuary context, domestic animals exclusively were used. Insofar as ancestral worship gave the credit to the kinship, the ownership of animals counted. The relationship between human and wild animals was relatively weak, whereas people could certainly declare ownership of their livestock. In addition, Fiskesjo also points out that the proximity of domestic animals to human beings made it possible to differentiate colour, age, sex and size in detail (*ibid.*). This would be less possible in the case of wild animals. Consequently, wild animals gradually became unsuitable as sacrificial offerings to ancestors. That is not to say that wild animals were not used in other ritual contexts at all such as feasting and rites pertaining to hunting. A distinction has to be made between hunting-related rituals and those offering of captured games to ancestors (Fiskesjo 2001). Probably, the ancestral sacrifice, as one representative of the *jia* 家 principle, was necessarily restricted to domestic animals.

Thus far, differentiation between the wild and the domestic in animal categorisation became discernable in the Shang dynasty. This contrast did not seem to synchronise the progress of the domesticating practice in which the time of the first appearance of morphologically domestic species (*i.e.* domestic dog at the Nanzhuangtou site) was some 6000 years earlier. Of course, people might have been aware of the differences brought to animals by domesticating processes. Such differences, however, were not firmly reified and manifested in the archaeological representation until the Shang dynasty, along with the expansion of *jia* 家 idea. The reinforcement of *jia* 家 principle in turn penetrated boundaries between the wild and the domestic, and hence forced the transformation between the two.

7.4 A glimpse beyond China

Though a meticulous comparative study seems unnecessary, it can helpfully provide a glimpse into animal categories beyond China, just go grant a general idea of what goings on elsewhere in the world.

During the third and first millennium BC, while China witnessed the rising of increasingly complex societies, on the other end of the continent, ancient Mesopotamia saw the emergence of the early kingdom, writing system, urban life and so on (Postgate 2015). Here animal taxonomy also displayed sharp contrast to the scientific taxonomy. Snake groups, for example, including lizard, turtle and fox, rather than being grouped into *Canidae* family, was seen as a category of "transition to the small cats" (Chalendar 2019). Studies into folk taxonomies in ancient Mesopotamia have proved fruitful, thanks to the written records available. Taking a more or less formalist approach, information has been retrieved from a number of different sources including trade documents, medical descripts, religious texts and so forth to reconstruct ancient categories.

Taking *materia medica,* or medical prescriptions for example, a relative intact taxonomic system can be reconstructed. There were some rules observed in prescriptions whereby all ingredients were categorised according to their affiliation with a natural domain (*i.e.* mineral, vegetal and animal domains), then arranged into subclasses according to their obvious features (*e.g.* aromatics and woods) and further into subcategories on account of their body parts (*e.g.* hair and tissues of animals.) (Chalendar 2019).

Medical plants and animals may form a distinctive category. In the region studied in this book, there were no traces indicating medical use of plant and remains during the research period. In Xinjiang, in northwest China, a cache of cannabis was preserved in a 2700-year-old grave at the Yanghai cemetery, due to the local environment being particularly good for preservation of organic remains (Jiang *et al.* 2006). The grave is believed to have belonged to a Caucasoid shaman and the cannabis presumably used as a medicinal or psychoactive agent (Russo *et al.* 2008). Stronger evidence for medical plants and animals were present in later historical periods in China, either recorded in historical texts or presented in archaeological contexts in most circumstances as grave goods. A medical book excavated from the Han dynasty Mawangdui cemetery included 115 medicinal plants and 57 medicinal animals (Liu 2016). More data will be necessary to push the study on medical plants and animals further back into prehistoric time.

The wild-versus-domestic tension has also been discussed in ancient contexts. The division has also been found in ancient Mesopotamia. In the Hittite world (*ca.* 1700~1200 BC), wild and domestic animals were explicitly separated into two groups. Wild animals included lion, wolf, fox, bear, snake, leopard, deer, gazelle and boar, whereas in the domestic category dog, pig, sheep, goat, bovine and equine were embraced (Peled 2019). Peled (2019) suggests that the boundaries between the two were strictly demarcated and breaches were forbidden. The result, as Mary Douglas suggests (1966), was taboos on consumption across boundaries in the Israelite view of things, as exemplified by the pig. This marks a sharp contrast to the wild-domestic division in China where in certain circumstances a boundary was constructed purposefully to be crossed.

Behind the arrangement of animal categories, a cosmological ideology was in operation. Again in the Hittite world, the hierarchy behind the organisation of animals was their relationship to the gods. Humans were always categorised in the higher classes, reflecting its superiority over animals and wild animals as numinous creatures were always superior to domestic ones (Peled 2019). The 'animals of the gods' in the Hittite world often included the leopard, lion, bear, boar, wolf, deer, gazelle and wild goat, all invariably wild (Noegel 2019). The animal world was a mirror reflecting how Hittites perceived their world. In this sense, the animal world was closely articulated to the more broad cosmos beyond, resembling the situation in ancient China.

7.5 Summary

This chapter synthesises the discussions from three archaeological sites and addresses animal classification in a wider cultural background. There are three main topics. The first section is devoted to discussing a definition of 'human' category. In the distinction from the purely biological definition, the category of 'human' is firmly situated in social identity back in the Neolithic and Bronze Age. The second topic concerns age-related classification. Age, as a classificatory filter, is influential in many aspects of life, from ritual ceremony to disposal of rubbish. The prevalence of the age classifier is also supported by lexical evidence. The last topic is on the domestic-versus-wild dichotomy. I rephrase the dichotomy in Chinese language as *jia/ye* 家 / 野 opposition in order to contextualise it within this specific research period and region. This contrast seems a complex process involving the interaction of political, economic, social and other domains. Constructed on the basis of the *jia/ye* 家 / 野 contrast, the structure, however, was not unmoving in its binary nature. While merits were asymmetrically biased towards the side of *jia* 家, efforts were also devoted to elaborating the wild domain as a reference for contrast. The wild/domestic differentiation formed gradually over time. Animal categories constituted one vital part of this broader classificatory scheme that concerned the organisation of the cosmos as a whole. Towards the end of the chapter, a glimpse into how people in other parts of the world classified animals proved useful towards rendering a comparative perspective.

8

Conclusions

This chapter addresses my conclusions from several aspects. The first set of conclusions recapitulates the methodology proposed in chapter two, in order to explore the correlation between taxonomy and deposition. The second set of conclusions relating the methodological application to the archaeological assemblages, restates the understanding of animal categories in central China during the late Neolithic and the early Bronze Age, as demonstrated in chapter four to seven. The third section revisits several research themes introduced in chapter one. This section also enumerates unexplored issues due to the restriction of our datasets. The final set of conclusions poses questions for future studies before our contribution to the present state of archaeology is summarise and we attempt to locate this project within its wider academic and social horizon.

8.1 Research methodology recapitulated

In chapter two, it is argued that taxonomy, beyond perceptual organisation of things, concerns a process involving both physical behaviour and materialised representation. To explore past taxonomies, contextual archaeology is integrated with structured deposition studies, so as to lay the theoretical foundation of this study. Contextual archaeology is refined in order to underline the contextualisation of archaeological assemblage. At the same time, by addressing the correlation between taxonomy and deposition, this study argues that the idea of structured deposition has a more general relevance than previously proposed. It is rather the intention than the structure that determines the nature of a deposition. As long as people dispose of things intentionally, depositional behaviour is always guided by taxonomic consideration and constitutes taxonomic action.

Within this theoretical framework, zooarchaeology, along with isotope analysis, spatial analysis and linguistic reference are brought together to establish a methodology. This approach aims to retrieve information from bones and associate bones with depositional contexts and then allow the information to represent the differentiation between treatments of animals from an emic perspective.

Having been applied to three archaeological sites in China, as shown in chapter four, five and six, this method serves as a useful tool to examine depositional processes and hence explore the animal classifications informed by depositions of bones. The exact analytical tools employed vary depending on the datasets available in these three sites, but the methodological core remains the same — following the thread of the evidence and integrating information from various lines of evidence.

8.2 Animal classification reinterpreted

Inferring from the three sites, a set of characteristics regarding animal classification in ancient China has been observed and discussed in chapter seven. Briefly stated, there are three traits.

1) A categorical filter relating to age. Age appeared to be a fairly pervasive categorical criterion discovered at all three sites and applied to both human and non-human animals. The binary classification modified by age depicted a sharp contrast between the young and the adult. On one hand, classification guided the practice of the lifeway in prehistoric central China. On the other hand, the practice itself in turn contributed to consolidating classificatory scheme through formalisation and repetition.

2) The wild-versus-domesticated division established in the Central Plains around the Shang dynasty. This dichotomy, though deeply imprinted on the modern mind, did not apparently occur simultaneously with the domestication of animals *qua* husbandry practice. The opposition had not been clearly expressed and valued seriously until the Shang dynasty, when both natural and social domains began to participate in formulating this division. Though during the Xinzhai period, the segregation between the 'inside' and the 'outside' was vaguely constructed as signaled by the erection of fortified walls, this separation was rather permeable, with their boundaries being fairly porous.

3) The integration of animal categories into a larger correlative classificatory scheme. This trait is partially derived from the connection between animals and spaces stated in the second one. It targeted a wholistic explanation of the cosmos. For example, age categories were employed in association with ritual performance at Wadian, while the wild-versus-domesticated division corresponded to spatial segregation at Xinzhai. This integration, consistently reinforced in later dynastic periods, as demonstrated in chapter one, had its roots in the Longshan period at the latest.

8.3 Research themes revisited

Chapter one proposes four main themes which then formed the main exploratory avenues for the project. Having explored archaeological methods for approaching taxonomy and discussed its application to the case of ancient China, it is time to recall these themes and refresh our understanding of them.

8.3.1 Etic and emic

This study principally emphasises an emic perspective, and thus a relativist approach is taken. The weak engagement of etic research strategy, which otherwise would have been suggestive of cross-cultural comparisons, inevitably results in the case-by-case specifications in the conduct of the research. One weakness brought by such specification is that the application of the approach heavily relies on precise context. This precision of information is pre-determined by a variety of factors, including preservation, excavation methods, accessibility to the original record and so on, all of which can hardly be manipulated by researchers. This situation requires the method adapt to datasets with diverse characteristics. Consequently the approach employed in this book, rather than being a set of static operations, demonstrates its own constant progress of adaptation and refinement. As seen in studies of the three sites, some analyses were not employed in certain case due to their incompatibility to the datasets. For example, the pig-versus-deer index was calculated for the Wadian site considering their predominate proportions in the assemblage. These indices indeed are straightforward to demonstrate the correlation between the selection of animals for deposition and the location of the deposition (see ch 4). This index, however, was not calculated in other sites, primarily because of the relative paucity of deer bones. This would be a logical reason to explain leaving the deer bones out of the index.

Tension between specification and generalisation is central to emic/etic discussion. To the beginning of this study, it is aimed to dissect the tension between the two. Many more questions emerged alongside this analysis, however, such as the debate over the employment of etic category versus emic category in archaeology. This debate is not new in archaeology. Typology, for example, an artificial classification for the researchers' convenience (Hayden 1984), has been long questioned its capability to honestly reflect the past. Archaeological deposition provides an alternative to examine the category of objects instead of typology. As exemplified by animal taxonomy in China, deposition is taxonomy. Perhaps an interesting project in the near future would employ emic categories to classify archaeological assemblages to explore the human past. Our understanding of the past may be radically rewritten under a different taxonomic paradigm. This will definitely lead to a more intense discussion over etics and emics, and a more difficult choice between universal generalisation and cultural specification, especially during the days when an increasing number of inter-regional projects are launched and scientific techniques are capable of providing extremely detailed information. How to mediate etic explanation and emic understanding thus will be a major task for archaeologists. To some extent, an etic research strategy is also emic to researchers. This paradox further adds the complexity to archaeologists' role in searching for the past and the exact way to fulfill this task is always open to dispute.

8.3.2 Language and action

Both language and action are informed by taxonomy. The case study in China demonstrates the plausibility of the archaeological approach to cope with ancient taxonomies. Besides, the correspondence in animal categories in ancient China (*i.e.* age categories) found between archaeological deposition and textual record supports the language/action correspondence. At the same time, archaeological counterparts cannot be found for substantial animal categories recorded in ancient texts. We cannot anticipate that archaeological depositions can render categorical information with fine resolution as those nomenclature recorded in ancient texts. It is therefore not surprising that taxonomies manifested in depositions do not entirely correspond to those expressed in language. The disconnect between language and action invokes more issues for exploration, for example whether taxonomies recorded by language are exactly the same as those represented by depositions.

Due to the scope of this study and the resolution of the datasets, the investigation into this issue has to be curtailed for the time being. One possible conjecture about the Chinese case could be that there might have been variation of categorical schemes among different social groups. Even one society could have contained multiple taxonomies employed by different groups of people for different purposes. Taxonomies recorded in textual passages in ancient China were most likely composed by educated individuals. At the same time there might have been considerable illiterate farmers and hunters practicing their own taxonomies which naturally went unrecovered, not least by themselves. As a result, failing to find an exact correspondence between encyclopedic categorical definitions recorded in ancient texts and empirical taxonomic knowledge preserved in archaeological depositions is hardly surprising. This argument again recalls the tension between formalism and relativism, the former emphasising categorical names, the latter empirical conductions. As different types of knowledge are brought into focus, discrepancies in the results should cause us little surprise. This can also be linked back to the etic/emic tension. In some ancient texts, it is claimed that taxonomies used by different social groups were recorded. In that case, an analogy between ancient recording of taxonomies used by illiterate people and our anthropological exploration of indigenous societies can be found. The educated writer of the past would have played the role of the researcher, whereas the recorded ancient social groups were the indigenous community for observation in anthropology. Subsequently the recording complied in those ancient texts is itself a compound of etic and emic constructions, making the linguistic approach to ancient taxonomy more complicated and implicit.

The resolution of current datasets, however, does not afford a biographic inspection to the level of the individual, at least in this study. Therefore the exploration of taxonomy with regard to the social status of their agency remains a blank space. Little has been explored concerning the role of observers and actors in etic and emic constructions that is reflected in ancient textual passages. Future work to explore the tension between language-represented taxonomy and deposition-represented taxonomy may better be framed within the field of historical archaeology where on one hand local records are available for reference and on the other hand, human behaviour can be inferred from archaeological evidence.

8.3.3 Domesticated and wild

The notion of 'domestication' is a social construct and thus subject to change depending on historic and social context. The idea of 'wildness' is formed on the opposite end of 'domestication'. Therefore the definition of 'wildness' constantly changes. Boundaries become porous and changeable. The results of this study support the argument, as exemplified by the case of Xinzhai (see ch 6). It is hard to draw a definite dividing line, hence some animals become difficult to categorise. There are many more modern cases, such as creatures in the zoo. Children visiting a zoo are imbued with knowledge about wildlife, the examples they see being, technically, animals under care. This paradox best illustrates the messy boundaries between the wild and the domesticated, and the complexity of their relationship. No wonders there is no one versatile definition capable of thoroughly and convincingly explaining 'domestication'. Perhaps the answer to the problem is simply that there is no such definition.

Though it remained unclarified, many archaeologists in fact are interested in the technical aspect of 'domestication'. In other words, 'domestication' is equivalent to 'domesticating practice'. This project displays the possibility to include the perceptual dimension as one criterion to define the meaning of 'domestication'. This definition highlights the indigenous understanding shared within particular cultural group instead of the general principle agreed upon by researchers. Much more remains to be explored through this theme, which unfortunately cannot be entirely covered in this book. A future effort would merit investigating the relationship between domestication as a widely employed practice in human subsistence and domestic category as an indigenous perception towards nature, culture and their interaction.

8.4 Future studies and intellectual reflections

This study is an innovative attempt to challenge a rarely touched topic on the archaeological exploration of folk taxonomy. Efforts have been made to justify a proper approach based on available data. It is at the same time fully acknowledged that the dataset is limited and the approach requires further elaboration. Therefore several major issues are recognised and may be worthy of future investigation.

8.4.1 Taphonomy

The first issue concerns taphonomy. Faunal remains excavated from archaeological sites have experienced alterations to various degrees due to different factors and this process of preservation, with no doubt, affects the information that archaeologists can extract from bones (Behrensmeyer & Kidwell 1985). Therefore, taphonomy, which is 'the study of the transition, in all details, of organics from the biosphere into the lithosphere or geological record' according to Lyman (1994: 1), needs to be carried out under scrutiny.

In that this study collated currently available published data, it has inevitably dealt with data of various resolutions. For those data with lower resolution, such as faunal remains excavated fairly long time ago and thus lacking detailed information, a thorough taphonomic assessment would be impossible. This would likely lead to misinterpretation as the preservation of bones, conditioned by post-mortem depositional process, recovery techniques, and other factors, determines what information can be extracted concerning past human behaviours. In this study, the focused distinction between ritual and utility contexts is fairly obvious even at a coarse resolution. This is partially because that the assemblage examined are constituted of mainly sizable bones from mid- and large-sized mammals (*e.g.* pig, cattle, sheep/goat and deer), and the contextual difference presents a sharp contrast.

Nonetheless if more diverse animal taxa are involved, a more thorough taphonomic assessment based on well collected data will be essential. Fish bones, for example, being small and fragile in general, were not always collected systematically in early excavations (Li 2014; Mo 2016). It is not surprising that the contribution of fish to both human diet, and social/cultural life was once under-estimated due to the scarcity of excavated data. No fish remains have been reported in studied sites, likely a combined result of collecting techniques and actual preservation. Nonetheless there are examples in the Middle Yangtze Valley during the late Neolithic where fish might have been differentially categorised in consumption and funeral contexts. At the Daxi site (*ca.* 4500~3300 BC.) in Chongqing, complete fish skeletons were deposited as grave goods in nine tombs. Some were placed next to the human corpse, alongside the arms, and some even in the mouth (Institute of Archaeology CASS 2010: 430). In addition to the significance in funeral practices, fish were likely one of the major compositions in human diet in this region, as illustrated by the many pits filled with fish bones found in a number of archaeological sites such as Zhongbaodai (National Cultural Heritage Administration Three Georges Team 2001) and Guanmiaoshan (Institute of Archaeology CASS 2017). At Daxi, bones from pigs, cattle, deer, tiger and other mammals along with fish, tortoises and shellfish, were also recovered in thesettlement area, suggesting a subsistence relying primarily on wild animals and supplemented by limited exploitation of domesticates (Yang & Ding 2000). The animals exploited display a diversity in taxa. It, however, remains unexplored how these animals including fish were classified and in particular what role fish might have played in different aspects of social life. To address these problems, a taphonomic assessment of various types of faunal remains is essential.

8.4.2 Inter-regional variation

The second problem that must be borne in mind concerns regional-specificity. The focus of the current study is central China. Expanding into a more broad geographical setting, we can expect to see a picture of rich diversity. The Lower Yangtze Valley illustrates an impressive contrast.

The late Neolithic communities in the study region, central China, could be described as pragmatists (Li 2009; He 2016), in particular compared with those in the Lower Yangtze Valley. As manifested in animal economy, these two regions adopted distinctive modes of animal exploration (Dong & Yuan 2020), probably associated with different perceptions of animals. Liangzhu, for example, is a late Neolithic culture in the Lower Yangtze Valley dating to 3300~3800 BC. During its heyday, Liangzhu became an influential culture in the Lower Yangtze Valley. With the emerging of institutional groups, a complex social network articulating various types of settlements and covering an extensive area gradually formed, among which the Liangzhu city (in modern Yuhang, Zhejiang Province) was likely a political, social, economic and religious centre (Renfrew & Liu 2018). One feature of the Liangzhu culture is the elaborately worked jades, some of which were carved with zoomorphic patterns rendering information to decipher the animal symbolism. Unlike the funeral ritual practiced in central China, where offering sacrificial animals was a strong tradition since the Yangshao cultural period, recoveries of animal offerings in ritual contexts were scarce along the Lower Yangtze. Instead, jades were frequently found in luxurious tombs. The contrasting patterns in the use of animals for funeral rituals agree with the distinctive husbandry modes observed in pig management in the two regions: while in the central China pig husbandry immediately intensified following early domestication around 7000 BC, in the Lower Yangtze, the limited exploitation of domestic pigs was only supplementary to the the utilisation of wild resources (Dong & Yuan 2020). The practical utilisation of animals might have been guided by human perception on these respective creatures. In other words, roughly contemporaneously, distinctive systems of animal classification might have been employed across China. It is thus necessary to expand future research into this untapped region.

8.4.3 Animal imagery

Thirdly, additional strands of evidence from various aspects will be key to future studies. As repeatedly emphasised in previous chapters, animal classifications in ancient China were embedded into a broader system of cosmology. That is to say, animal taxonomies may also be manifested in various medias, in addition to the animals themselves. One of the medias was zoomorphic pattern on ritual objects including bronze, jade, stone and ceramic artifacts, as briefly illustrated in chapter four. The zoomorphic expression, a transformation of real animal shapes, could have been a further elaboration of the human perception of animals. The Shang, for example, featuring its own complicated ritual use of animals (Yuan & Flad 2005; Li & Campbell 2019) and magnificent bronze vessels with zoomorphic motifs (Chang 1981; Kesner 1991), merits further study. It was in the Shang dynasty that older Neolithic visual traditions (previously expressed in jades and ceramics) were incorporated into the aesthetic-symbolic ideology underscoring the Shang socio-political order (Kesner 1991). Apart from the visual language of animals carried by various medias from bronzes to jades, the bureaucratic ritual involving sacrificial animals also deliver the statements of Shang social hierarchy and political dominance. The two aspects pertaining to animals together intensified the Shang socio-political order. Only a few of studies attempt to explore the the potential connection between the animal art and the animals themselves integrating perspectives of art history and zooarchaeology more or less (*e.g.* Hayashi 1983; Campbell 2015), let alone the investigation into the prehistoric root of the animal taxonomy, which might have fostered both the animal economy and the animal symbolism expressed in visual art.

During the late Neolithic, despite of those reports about findings of visual expressions of animals in various forms in central China (*e.g.* Institute of Archaeology CASS & LMBC 2015; HPIACR & PMACR 2017), only a handful of studies of zoomorphic decorations systematically discuss the ideology behind those visual expressions (*e.g.* Chang 1981; Feng 1990). It has been noted in chapter four that the methodologies employed in art history and zooarchaeology are incommensurate. Besides, zoomorphic patterns, more than often, depict images of 'fantastic beasts' which lie beyond the scope of zooarchaeological research that focus primarily on real animals, widening the gap between art history and zooarchaeological studies. As a result, animal imagery is not discussed in this study but worth investigating in the future.

8.4.4 Integrating plants and animal remains

It has been repeatedly underscored that animal categories in ancient China were embedded in a nebulous cosmology embracing various elements. An isolated investigation into animals would not be sufficient to render a holistic view. Plant remains are as closely associated with animal remains especially in the research of ancient subsistence. Plants are also part of the folk taxonomy of the natural world. Studying two fundamentally incommensurate types of the assemblages remains problematic in many ways (Miller 2013). In future studies, it is hoped that archaeobotanical research will pay attention to the classification of ancient plants, thus adding another dimension for discussion.

Current archaeobotanical research in central China demonstrates agricultural production in the late Neolithic roughly relying on two major staple foods — foxtail millet (*Setaria italica*) and broomcorn millet (*Panicum miliaceum*) — supplemented by rice (*Oryza sativa*) and soybean (*Glycine max*) (Lee *et al.* 2007; Deng *et al.* 2015; Deng & Qin 2017). Wheat (*Triticun spp.*), a crop of southwest Asian origin, was brought into China by at least 2500 BC and began to be gradually incorporated into the local agricultural economy thereafter (Zhao 2009; Jones *et al.* 2011). By the Shang dynasty, wheat cultivation was drastically intensified in central China and began to play a potent role particularly among the elites in ritual activities (Chen 2016). Chen (2016) further speculates that the intensification of wheat cultivation was a top-down process stimulated by the Shang elites, one result of which was the fact that in regions outside the metropolis, such as eastern China, with its more suitable environment, wheat cultivation only expanded at a much slower pace. New radio-carbon dating results suggest that wheat might have not played a major subsistence role in central China until the Zhou dynasty, after 3000 BP (Deng *et al.* 2020).

The case of wheat provokes its own intriguing issues pertaining to folk taxonomy. It may provide clues to address how exotic crops were classified when they were brought into a new land and then incorporated into local agricultural system. Historians, dwelling on oracle bone inscriptions, have long attempted to justify the link between the categorical name of the wheat 来 (lai) meaning 'coming' and its implication for the introduction of this southwest Asian originated crop (Ho 1969, 1977; Fan 2002). The shifting role of wheat from a prestige crop privileged by the Shang elites towards a subsistence staple food in the Zhou dynasty very likely corresponded to a change in its category. Therefore an investigation into the patterning of archaeological data may cast light upon this issue. If the classification of wheat can be approached, it will be intriguing for the mirroring question in zooarchaeology: that is how those west Asian originated livestock such as sheep, goats and cattle were categorised and fit into an existing local taxonomy. A similar question has been addressed in an interdisciplinary study of the animal categories *nahiru* in Akkadian texts. By tracing the process of the lexical change, Wapnish (1995) demonstrates that the Akkadians borrowed the name of the animal they were familiar with to name unfamiliar new comers which very likely were a group of sea mammals sharing similar characteristics, rather than a particular species of animal. As in the case of China, both west Asian originated crops and livestock were introduced before the time of written records. Archaeology hence may render an alternative approach.

8.4.5 Past for the present

To sum up, this study firstly contributes a proposal of employing folk taxonomy to approach human-animal relationship in the past. Secondly, it develops an archaeological methodology to approach folk taxonomies used by people in the past through material representation. Thirdly, the application of this methodology casts light upon indigenous classifications of animals in ancient China.

As proposed at the outset, the tension between Linnaean and folk taxonomies mirrors the tension between etic and emic approaches. It is from these tensions that this study initially began to take shape. The study, simply stated, aims to take an emic perspective to approach animal categories in the past, rather than simply project the grid of Linnaean classification onto animals of the past. Beyond the contribution towards the methodological construction and the understanding of animal categories in one particular time period and society, what is more important is that the project alert us to two risks threatening our research of the past: anachronism and teleology. These two dangers together may lead to the distortion of the past under our own epistemology and ideology. By no means should Linnaean taxonomy be entirely excluded. The book attests how Linnaean system remains an invaluable basis for recording and comparison. Archaeology informs us how to address the heterogeneity of the past and thus the present.

The category of 'human' is, for example, not only an issue about the past, but also a heated debate in modern societies, as in issues about determining the start of life and its consequence on the legitimation of abortion. When does a 'human' begin the life? Is the answer at conception, birth, or when he or she becomes an 'adult', as demonstrated in the cases of Wangchenggang and Xinzhai? If the Wanchenggang and Xinzhai societies did not treat children as proper 'humans', does it mean that we are more 'human' than people in the past? In fact, this question itself is incommensurably problematic to me, as it exercises modern judgment of the past activities. The point that I make here is that the past teaches us of a world of diversity. Heterogeneity is intrinsic to the past, the present and the future. Understanding a heterogonous world involves placing it back in context and understand its own rationale.The principle is not only applied to societies in the past but also societies today. An understanding of others should not be clouded by our own ideology.

I hope this book cannot only enrich our understanding of animal categories in ancient China but also contribute to the intellectual reflection on the connection between the past and the present made through archaeology.

Bibliography

ACSACPU (Aurora Centre for the Study of Ancient Civilization, Peiking University and ZMIACR (Zhengzhou Municipal Institute of Archaeology and Cultural Relics). 2008. *Xinmi Xinzhai (Excavation of Xinzhai)*. Beijing: Wenwu Press. (in Chinese)

Albarella, U., K. Dobney, and P. Rowley-Conwy. 2006. "The domestication of the pig (*Sus scrofa*): new challenges and approaches." In *Documenting Domestication: New Genetic and Archaeological Paradigms,* edited by M. Zeder, 209–27. Berkeley: University of California Press.

Allan, S. 1991. *The Shape of the Turtle : Myth, Art, and Cosmos in Early China.* New York: State University of New York Press.

An, J. 1985. "Shi lun Dengfeng Wangchenggang Longshan wenhua chengzhi yu Xia Dai Yangcheng (On the Wangchenggang site and the Yangcheng site)." In *Zhongguo Kaoguxue Disici Nianhui Lunwenji (Proccedings of the 4th Annual Meeting of Chinese Archaeology)*, edited by Chinese Archaeology Association, 1–6. Beijing: Wenwu Press. (in Chinese)

Anderson, K. 1997. "A walk on the wild side: a critical geography of domestication." *Progress in human geography* 21 (4): 463–85.

Atkinson, Q. and H. Whitehouse. 2011. "The cultural morphospace of ritual form: Examining modes of religiosity cross-culturally." *Evolution and Human Behavior* 32(1): 50–62.

Atran, S. 1990. *Cognitive Foundations of Natural History: Towards an Anthropology of Science.* Cambridge: Cambridge University Press.

— 2001. "Folkbiology." In *The MIT Encyclopedia of the Cognitive Sciences*, edited by R. Wilson and F. Keil, 316–7. New York: MIT Press.

ATYRR (Archaeologcial Team for Yellow River Reservoir). 1960. "Henan Shanxian Qilipu Shangdai yizhi de fajue (Excavation of the Qilipu site)." *Kaogu Xuebao* (1): 25–49. (in Chinese)

Bailey, G. 2007. "Time perspectives, palimpsests and the archaeology of time." *Journal of Anthropological Archaeology* 26 (2): 198–223.

Barnard, A. 2002. "Emic and etic." In *Encyclopedia of Social and Cultural Anthropology*, edited by A. Barnard and J. Spencer, 275–8. London: Routledge.

Barrett, J. 1987. "Contextual archaeology." *Antiquity* 61: 468–73.

Barton, L., S. Newsome. and F. Chen. 2009. "Agricultural origins and the isotopic identity of domestication in northern China." *Proceedings of the National Academy of Sciences* 106 (14): 5523–28.

Behrensmeyer, A. K. and S. M. Kidwell. 1985. "Taphonomy's contributions to paleobiology." *Paleobiology* 11: 105–19.

Benvenistes, E. and J. Lallot. 1969. *Le Vocabulaire des Institutions Indo-Europeaenes 1*. Paris: Les Editions de Minuit.

Berlin, B. 1972. "Speculations on the growth of ethnobotanical nomenclature." *Language in Society* 1 (1): 51–86.

— 1973. "Folk systematics in relation to biological classification and nomenclature." *Annual Review of Ecology and Systematics* 4: 259–71.

— 1974. "Further notes on covert categories and folk taxonomies: a reply to Brown." *American Anthropologist* 76 (2): 327–31.

— 1976. "The concept of rank in ethnobiological classification: some evidence from Aguaruna folk botany." *American Ethnologist* 3: 381–99.

Berlin, B., D. Breedlove. and P. Raven. 1966. "Folk taxonomies and biological classification." *Science* 154 (3746): 273–5.

— 1973. "General principles of classification and nomenclature in folk biology." *American Anthropologist* 75 (1): 214–42.

Berry, J. 1989. "Imposed etics-emics-derived etics: the operationalization of a compelling idea." *International Journal of Psychology* 24 (6): 721–35.

Binford, L. 1980. "Willow smoke and dogs' tails: hunter-gatherer settlement systems and archaeological site formation." *American Antiquity* 45 (1): 4–20.

Blanton, R. 1994. *Houses and Households: A Comparative Study*. New York: Plenum.

Boyle, K. 2006. "Neolithic wild game animals in Western Europe: the question of hunting." In *Animals in the Neolithic of Britain and Europe*, edited by D. Serjeantson and D. Field, 10–23. Oxford: Oxbow Books.

Bradley, R. 1990. *The Passage of Arms: an Archaeological Analysis of Prehistoric Hoards and Votive Deposits*. Cambridge: Cambridge University Press.

Breedlove, B. and P. Raven. 1968. "Covert categories and folk taxonomies." *American Anthropologist* 70 (2): 290–9.

Broda, J. 1991. "The sacred landscape of Aztec calendar festivals: myth, nature, and society." In *To Change Place: Aztec Ceremonial Landscapes*, edited by D. Carrasco, 74–120. Niwot: University Press of Colorado.

Browne, E. J. 1983. *The Secular Ark : Studies in the History of Biogeography*. New Heaven: Yale University Press.

Bruck, J. 1995. "A place for the dead: the role of human remains in Late Bronze Age Britain." *Proceedings of the Prehistoric Society* 61: 245–77.

— 1999. "Ritual and rationality: some problems of interpretation in European archaeology." *European Journal of Archaeology* 2 (3): 313–44.

Brunson, K., R. Lele, Z. Xin, D. Xiaoling, W. Hui, Z. Jing and R. Flad. 2020. "Zooarchaeology, ancient mtDNA, and radiocarbon dating provide new evidence for the emergence of domestic cattle and caprines in the Tao River Valley of Gansu Province, northwest China." *Journal of Archaeological Science: Reports* 31. 102262.

Buck, C. 1949. *A Dictionary of Selected Synonyms in the Principal Indo-European Languages*. Chicago: University of Chicago Press.

Bulmer, R. 1967. "Why is the cassowary not a bird? A problem of zoological taxonomy among the Karam of the New Guinea Highlands." *Man* 2 (1): 2–25.

Cai, D., Z. Tang, L. Han, C. Speller and D. Yang. 2009. "Ancient DNA provides new insights into the origin of the Chinese domestic horse." *Journal of Archaeological Science* 36 (3): 835–42.

Cai, D., Z. Tang, H. Yu, L. Han, X. Ren and X. Zhao. 2011. "Early history of Chinese domestic sheep indicated by ancient DNA analysis of Bronze Age individuals." *Journal of Archaeological Science* 38 (4): 896–902.

Cai, D., L. Han, X. Zhang, H. Zhou. and H. Zhu. 2007. "DNA analysis of archaeological sheep remains from China." *Journal of Archaeological Science* 34 (9): 1347–55.

Cai, X., H. Chen, C. Lei, S. Wang, K. Xue and B. Zhang. 2007. "MtDNA Diversity and genetic lineages of eighteen cattle breeds from *Bos taurus* and *Bos indicus* in China." *Genetica* 131 (2): 175–83.

Campbell, R., Z. Li, Y. He and Y. Jing. 2011. "Consumption, exchange and production at the Great Settlement Shang: bone-working at Tiesanlu, Anyang." *Antiquity* 85 (330): 1279–97.

Campbell, R. 2009. "Toward a networks and boundaries approach to early complex polities." *Current Anthropology* 50 (6): 821–48.2015

— 2015. "Animal, human, god: pathways of Shang animality and divinity." In *Animals and Inequality in the Ancient World*, edited by B. Arbuckle and S. McCarty, 251–74. Boulder: University Press of Colorado.

Cao, G. and Q. Ma. 1983. "Henan huaiyang Pingliangtai Longshan wenhua chengzhi shijue jianbao (Excavation report of the Pingliangtai site)." *Wenwu* (3): 21–36. (in Chinese)

Cao, J. 2008. "Zhou dai jisi yongsheng lizhi kaolue (A brief study on the ritual system of animal sacrifice in Zhou dynasty)." *Wenbo* (3): 18–21. (in Chinese)

Carden, R. F. 2006. "*Putting Flesh on Bones: The Life and Death of the Giant Irish Deer (Megaloceros giganteus, Blumenbach, 1803)*." PhD diss., University College of Dublin.

Chalendar, V. 2019. "Taxonomy and medicine." In *Animals and their Relations to Gods, Humans and Things in the Ancient World*, edited by R. Mattila, S. Ito and S. Fink, 59–78. Wiesbaden: Springer VS.

Chan, K. 1981. "The animal in Shang and Chou bronze art." *Harvard Journal of Asiatic Studies* 41 (2): 527–54.

— 1983. "Settlement patterns in Chinese archaeology: a case study from the Bronze Age." In *Prehistoric Settlement Patterns: Essays in Honor of Gordon R. Willey*, edited by E. Z. Vogot and R.M. Leventhal, 361–74. Albuquerque: University of New Mexico Press.

Chapman, J. 2000. "Pit-digging and structured deposition in the Neolithic and Copper Age." *Proceedings of the Prehistoric Society* 61: 61–7.

Chen, G., Q. Cai, J. Ma, Y. Li, H. Yu. and Q. Liu. 2002. "Henan Xinmishi Guchengzhai Longshan wenhua chengzhi shijue baogao (Excavation of the Longshan city Guchengzhai in Henan Province)." *Huaxia Kaogu* (2): 53–82. (in Chinese)

Chen, W. 1994. *Zhongguo Nongye kaogu Tulu (Agricultural Archaeology Catalogue of China)*. Jiangxi: Jiangxi Scientific and Technical Publishers. (in Chinese)

Chen, X. 2016. "Zhongguo qingtong shidai xiaomai zhongzhi guimo de kaoguxue guancha (An archaeological estimation of the scale of wheat cultivation in Bronze Age China)." *Zhongguo Nongshi (3)*: 3–9.

Chen, X., Y. Li. and L. Liu. 2010. "2002-2003nian Henan Yanshi Huizui yizhi de fajue (The excavation at the Huizui site during 2002-2003)." *Kaogu Xuebao* (3): 1–7. (in Chinese)

Chen, X., Y. Fang, Y. Hu. and Y. Hou. 2015. "Isotopic reconstruction of the late Longshan period (*ca.* 4200–3900 BP) dietary complexity before the onset of state-level societies at the Wadian site in the Ying River Valley, Central Plains, China." *International Journal of Osteoarchaeology* 26: 808–17.

Childe, V. G. 1950. "The urban revolution." *Town Planning Review* 21 (1): 3–17.

— 1957. *The Dawn of European Civilization.* London: Rutledge & Paul.

Chinese Archaeology. 2013 "Wangchenggang online faunal dataset." http://www.kaogu.cn/cn/xueshuziliao/kaogushujuku/dongwukaoguziliaoku/2013/1025/31716.html

Clarke, S. 1997. "Abandonment, rubbish disposal and special deposits at Newstead." In *TRAC 96: Proceedings of the Fourth Annual Theoretical Roman Archaeology Conference 1996,* edited by K. Meadows, C. Lemke, and J. Heron, 73–81. Oxford: Oxbow Books.

Clutton-Brock, J. 1995. "Origins of the dog: domestication and early history." In *The Domestic Dog: Its Evolution Behavior and Interaction with People*, edited by J. Serpell, 7–20. Cambridge: Cambridge University Press.

Clutton-Brock, T., F. Guinness. and S. Albon. 1982. *Red Deer: Behavior and Ecology of Two Sexes*. Edinburgh: Edinburgh University Press.

Coblin, W. 1993. "Erya." In *Early Chinese Texts: A Bibliographical Guilde*, edited by M. Loewe, 94–9. Berkeley: Study of Early China : Institute of East Asian Studies, University of California, Berkeley.

Conklin, H.C. 1954. "*The Relation of the Hanunoo Culture to the Plant World.*" PhD diss., Yale University.

— 1962. "Lexicographical treatment of folk taxonomies." In *Problems of Lexicography*, edited by F. W. Householder and S. Saporta. Indiana University Research Center in Anthropology and Human Behavior & Anthropological Society of Washington.

Conover, M. R. 2001. *Resolving Human-Wildlife Conflicts: The Science of Wildlife Damage Management.* New York: Lewis Publishers.

Cucchi, T., L. Dai, M. Balasse, C. Zhao, J. Gao and Y. Hu. 2016. "Social complexification and pig (*Sus scrufa*) husbandry in ancient China: a combined geometric morphometric and isotopic approach." *PloS one* 11 (7): e0158523.

Cucchi, T., A. Hulme-Beaman and J. Yuan. 2011. "Early Neolithic pig domestication at Jiahu, Henan Province, China: clues from molar shape analyses using geometric morphometric approaches." *Journal of Archaeological Science* 38 (1): 11–22.

Dai, L., M. Balasse, J. Yuan, C. Zhao and Y. Hu. 2016."Cattle and sheep raising and millet growing in the Longshan age in central China: Stable isotope investigation at the Xinzhai site." *Quaternary International* 426: 145–57.

Dai, L., Z. Li., Y. Hu, C. Zhao and C. Wang. 2014. "Xinzhai yizhi chutu yang de siwang nianling ji chuchanpin kaifa celue (Sheep secondary products exploration at the Xinzhai site)." *Kaogu* (1): 94–103. (in Chinese)

Dai, L., Z. Li, C. Zhao, J. Yuan and L. Hou. 2015. "An isotopic perspective on animal husbandry at the Xinzhai site during the initial stage of the legendary Xia Dynasty (2070–1600 BC)." *International Journal of Osteoarchaeology* 26 (5): 885–96.

Douglas, M. 1966. *Purity and Danger: an Analysis of Concepts of Pollution and Taboo.* London: Routledge and Paul.

De Chardin, T.P. and C.C. Young. 1936. *On the Mammalian Remains from the Archaeological Site of Anyang.* Nanking: Geological Survey of China.

Dematte, P. 2006. "The Chinese Jade Age Between antiquarianism and archaeology." *Journal of Social Archaeology* 6 (2): 202–26.

Deng, Z. and L. Qin. 2017. "Zhongyuan Longshan shidai nongye jiegou de bijiao yanjiu (A comparative study of agricultural structure in the central plains during the Longshan period)." *Huaxia Kaogu 3*: 98–108. (in Chinese)

Deng, Z., D. Q. Fuller, X. Chu, Y. Cao, Y. Jiang, L. Wang and H. Lu. 2020. "Assessing the occurrence and status of wheat in late Neolithic central China: the importance of direct AMS radiocarbon dates from Xiazhai." *Vegetation History and Archaeobotany* 29: 61–73.

Deng, Z., L. Qin, Y. Gao, A.R. Weisskopf, C. Zhang and D.Q. Fuller. 2015. "From early domesticated rice of the middle yangtze basin to millet, rice and wheat agriculture: archaeobotanical macro-remains from Baligang, Nanyang Basin, Central China (6700–500 BC)." *PLOS ONE* 10 (10): e0139885.

Dentan, R. 1970. "Labels and rituals in Semai classification." *Ethnology* 9 (1): 16–25.

Dobney, K. and K. Rielly. 1988. "A method for recording archaeological animal bones: the use of diagnostic zones." *Circaea* 5 (2): 79–96.

Dong, N. and J. Yuan. 2020. "Rethinking pig domestication in China: regional trajectories in central China and the Lower Yangtze Valley." *Antiquity 94* (376): 864–79.

Dorofeeva-Lichtmann, V. 1995. "Conception of Terrestrial Organization in the Shan hai jing." *Bulletin de l'Ecole française d'Extrême-Orient* 82: 57–110.

Driver, J.C. 2011. "Identification, classification and zooarchaeology." *Ethnobiology letters* 2: 19–39.

Duan, P., Y. Du and H. Xiao. 1984. "Yanshi Shangcheng de chubu kantan he fajue (Preliminary survey and excavation at the Yanshi site)." *Kaogu* (6): 488–504. (in Chinese)

Eliade, M. and W. Trask. 1958. *Rites and Symbols of Initiation.* New York: Harper and Row.

Ellen, R. 1999. "Models of subsistence and ethnobiological knowledge: between extraction and cultivation in Southeast Asia." *Folkbiology* 1: 91–117.

— 2006. *The Categorical Impulse: Essays in the Anthropology of Classifying Behaviour.* Oxford: Bergham Books.

— 2008. "Ethnomycology among the Nuaulu of the Moluccas: putting Berlin's 'general principles' of ethnobiological classification to the test." *Economic Botany* 62 (3): 483–96.

Falkenhausen, L. Von. 1993. "On the historiographical orientation of Chinese archaeology." *Antiquity* 67 (257): 839–49.

Fan, Y. 2002. "Guanyu Shang dai mailei zuowu de jige wenti (Discussion on the wheat crops in Shang dyansty)." *Zhongguo Nongshi* 21 (1): 23–26. (in Chinese)

Fang, Y. 2002. "Settlement patterns in the upper Ying river valley based on excavations at the Wadian site." *Seventeenth Congress of the Indo-Pacific Prehistory Association*: 9–15. Taipei.

FAO, (Food and Agriculture Organization of the United Nations). 1982. Deer farming guidelines on practical aspects. Retrieved on 12-12-2016 from, http://www.fao.org/docrep/004/X6529E/X6529E08.htm

Feng, S. 1990. "Henan Puyang Xishuipo 45hao mu de tianwenxue yanjiu (The astronological meaning of the tomb M45 at the Xishuipo site, Puyang, Henan)." *Wenwu* (3): 52–6. (in Chinese)

Fiennes, C. 1888. *Through England on a Side Saddle in the Times of William and Mary. Being the Diary of Celia Fiennes. With an Introduction by the Hon. Mrs. Griffiths. London.* London: Field & Tuer.

Fiskesjo, M. 1999. "On the 'raw' and the 'cooked' barbarians of imperial China." *Inner Asia* 1 (2): 139–68.

— 2001. "Rising from blood-stained fields: Royal hunting and state formation in Shang China." *Bulletin-Museum of Far Eastern Antiquities* (73): 48–191.

— 2012. "The animal other: China's barbarians and their renaming in the 20th Century." *Social Text* 29 (4): 57–79.

Flad, R.K., J. Yuan. and S. Li. 2007. "Zooarcheological evidence for animal domestication in northwest China." In *Development in Quaternary Sciences vol.9,* edited by D. B. Madsen, F. Chen, and X. Gao, 167–203. Elsevier Science.

Fletcher, J. 2000. *A Life for Deer : A Deer Vet Tells His Story and Theirs*. London: Gollancz in association with Peter Crawley.

Fried. MH. 1960. *On the Evolution of Social Stratification and the State*. Indianapolis: Bobbs-Merrill.

Gao, G. and W. Shao. 1986. "Zhongguo shiqian shidai de guiling yu quansheng (Tortoise spirit and dog sacrifice in Chinese prehistory)." In *Zhongguo Kaoguxue Yanjiu (Chinese Archaeological Studies)*, edited by Editorial Board of Chinese Archaeological Study, 57–70. Beijing: Wenwu Press. (in Chinese)

Gao, J. 2005. "Xinzhai yizhi yu Zhongguo gudai wenming qiyuan wenti (The Xinzhai site and the origin of Chinese civilization)." *Zhongyuan Wenwu* (4): 22–7. (in Chinese)

Garrow, D. 2012. "Odd deposits and average practice: a critical history of the concept of structured deposition." *Archaeological Dialogues* 19: 85–115.

Gimbutas, M. 1952. "On the origin of North Indo-Europeans." *American Anthropologist* 54 (4): 602–11.

— 1970. *Proto-Indo-European Culture: the Kurgan Culture during the Fifth, Fourth, and Third Millennia BC*. Philadelphia: s.n.

Goodenough, W.H. 1970. *Description and Comparison in Cultural Anthropology*. Chicago: Aldine Publishing Company.

Grant, A. 1975. "Appendix B: The use of tooth wear as a guide to the age of domestic animals – a brief explanation." In *Excavations at Portchester Castle 1*, edited by B. Cunliffe, 450. London: Society of Antiquaries.

— 1982. "The use of tooth wear as a guide to the age of domestic ungulates." In *Ageing and Sexing Animal Bones from Archaeological Sites*, edited by B. Wilson, C. Grigson, and S. Payne, 91–108. Oxford: BAR Publishing British Series 109.

— 1984. "Survival or sacrifice? A critical appraisal of animal burials in Britain in the Iron Age." In *Animals in Archaeology*, edited by C. Grigson and J. Clutton-Brock, 221–7. Oxford: BAR Publishing International Series 227(4).

Grayson, D. 1981. "The effects of sample size on some derived measures in vertebrate faunal analysis." *Journal of Archaeological Science* 8 (1): 77–81.

— 1984. *Quantitative Zooarchaeology*. New York: Acadamic Press.

Greenberg, L. 1992. "Garden hunting among the Yucatec Maya: a coevolutionary history of wildlife and culture." *Etnoecológica* 1 (1): 23–33.

Gu, W. 2002. "Xinzhai qi yanjiu (On the Xinzhai phase)." *Yingdu Xuekan* (2): 26–40. (in Chinese)

Gu, W. and S. Zhang. 2003. "Gongyi Huadizui yizhi faxian 'Xinzhai'qi yicun (Discovery of the Xinzhai culture at Huadizui site)." *Gudai Wenming Yanjiu Tongxun 18 (News of Ancient Civilization Studies 18)*. Beijing: ACSAC. (in Chinese)

Guo, M. 1982. *Guo Moruo Quanji: Lishi Bian vol.1 (Works of Guo Moruo: History Part vol.1)*. Beijing: Renmin Press. (in Chinese)

Guo, G., Q. Cheng and J. Needham. 1999. *Zhongguo Gudai Dongwuxue Shi (The History of Zoology in Ancient China)*. Beijing: Kexue Press. (in Chinese)

Guo, J., L. Du, Y. Ma, W. Guan. and H. Li. 2005. "A novel maternal lineage revealed in sheep (*Ovis aries*)." *Animal Genetics* 36 (4): 331–36.

Habermehl, K. 1975. *Die Altersbestimmung bei Haus-und Labortieren*. Berlin: Paul Parey.

Hamilakis, Y. 2003. "The sacred geography of hunting: wild animals, social power and gender in early farming societies." *Zooarchaeology in Greece: recent advances* 9: 239–47.

Hamilton, J., R. E. Hedges and M. Robinson. 2009. "Rooting for pigfruit: pig feeding in Neolithic and Iron Age Britain compared." *Antiquity* 83 (322): 998–1011.

Hammer, Ø., D.A.T. Harper. and P.D. Ryan. 2001. "PAST: Paleontological Statistics Software Package for education and data analysis." *Palaeontolia Electronica* 4 (1): 9.

Harris, M. 1976. "History and significance of the emic/ etic distinction." *Annual review of anthropology* 5 (1): 329–50.

Hayashi, M. 1983. "In Sei-Shu jidai no dobutsu isho ni torareta yasei dobutsu rokushu (Six wild animals illustrated zoomorphic patterns during the Shang and Western Zhou dynasties)." In *Tembo Aijia no Kokogaku: Higuchi Takayasu Kyoju Taikan Kinen Roshu (The Future of Asian Archaeology: Essays in Honour of Higuchi Takayasu)*, edited by T. Higuchi, 549–52. Tokyo: Shinchosha. (in Japanese)

Hayden, B. 1984. "Are emic types relevant to archaeology?" *Ethnohistory* 31 (2): 79–92.

Hawkins, M. 1997. *Social Darwinism in European and American thought, 1860-1945*. Cambridge: Cambridge University Press.

He, N. 2016. "Shiqian jingji jichu kaogu shiye li de zuichu zhongguo xingcheng (The formation of early China from the economic perspective)." *Nanfang Wenwu* 2: 20–7. (in Chinese)

Hibbett, D. and J. Taylor. 2013. "Fungal systematics: is a new age of enlightenment at hand?" *Nature Reviews Microbiology* 11 (2): 129–33.

Hill, J.D. 1995. *Ritual and Rubbish in the Iron Age of Wessex: A Study on the Formation of a Specific Archaeological Record*. Oxford: Oxford BAR.

Hill, J. N. and R. K. Evans. 1972. "A model for classification and typology." In *Models in Archaeology*, edited by D. Clarke, 231–73. Oxon: Routledge.

Ho, C. W. 2009. *Windows on the Chinese World : Reflections by Five Historians*. Lanham: Lexington Books.

Ho, P. 1969. "The Loess and the Origin of Chinese Agriculture." *The American Historical Review* 75: 1–36.

— 1977. "The indigenous origins of Chinese agriculture." In *Origins of Agriculture*, edited by C.A. Reed, 413-83–58. The Hague: Mouton.

Hodder, I. 1982a. *Symbols in Action: Ethnoarchaeological Studies of Material Culture*. Cambridge: Cambridge University Press.

— 1982b. *The Present Past: An Introduction to Anthropology for Archaeologists*. London: B.T. Batsford Ltd.

— 1984. "Burials, houses, women and men in the European Neolithic." In *Ideology, Power and Prehistory*, edited by D. Miller and C. Tilley, 51–58. Cambridge: Cambridge University Press.

— 1990a. "Style as historical quality." In *The Uses of Style in Archaeology*, edited by M. Conley and C.A. Hastorf, 44–51. Cambridge: Cambridge University Press.

— 1990b. *The Domestication of Europe : Structure and Contingency in Neolithic Societies*. Oxford: Blackwell Publishing Ltd.

— 1994. "Theoretical archaeology: a reactionary view." In *Interpreting Objects and Collections*, edited by S. Pearce, 48–52. London: Routledge.

Hodder, I. and C. Cessford. 2004. "Daily practice and social memory at Çatalhoyuk." *American Antiquity* 69 (1): 17–40.

Hopwood, T. 1959. "The development of pre-Linnaean taxonomy." *Proceedings of the Linnaean Society of London* 170 (3): 230–34.

HPIACR, (Henan Provincial Institute of Archaeology and Cultural Relics). 1958. "Zhengzhou Niuzhai Longshan wenhua yizhi fajue baogao (A brief report of the Longshan period Niuzhai site)." *Kaogu Xuebao* (4). (in Chinese)

— 1983. "Zhengzhou Erqilu xin faxian sanzuo shangmu (Three Shang tombs found at Erqilu, Zhengzhou city)." *Wenwu* (3): 60–77. (in Chinese)

— 1989. "Zhenzhou shangdai Erligang qi zhutong jizhi (Metal foundries in the Zhenzhou site during the Erligang period)." In *Kaoguxue Jikan (6) (Essays on Archaeology, vol. 6)*, edited by Kaogu Press: 100–122. Beijing: Kaogu Press. (in Chinese)

— 1990. "Henan Jiashi Huizui yizhi fajue jianbao (A brief report of the excavation at the Huizui site)." *Huaxia Kaogu* (1): 1–33. (in Chinese)

— 1993a. "Henan Mixian Huangzhai yizhi de fajue (Excavation at the Huangzhai site, Henan Province)." *Huaxia Kaogu* (3). (in Chinese)

— 1993b. "Henan Gongxian Shaochai yizhi fajue baogao (A brief excavation report of the Shaochai site)." *Huaxia Kaogu* (2): 1–45. (in Chinese)

— 1996. "Henan Xingyang Shuhe yizhi fajue baogao (Excavation report of the Shuhe site, Henan Province)." In *Kaoguxue Jikan (10) (Essays on Archaeology, vol. 10)*, edited by Z. Wang. Beijing: Dizhi Press. (in Chinese)

— 1999. *Wuyang Jiahu (Excavation of the Jiahu site)*. Beijing: Kexue Press. (in Chinese)

— 2002. "Henan Xinmishi Guchengzhai Longshan wenhua chengzhi fajue jianbao (Brief excavation report of the walled Longshan site at Guchengzhai in Xinmi, Henan)." *Huaxia Kaogu* (2): 53–82. (in Chinese)

— 2004. *Yuzhou Wadian (Excavation of Wadian)*. Beijing: Shijie Tushu Press. (in Chinese)

— 2012. *Yancheng Haojiatai (Excavation of Haojiatai)*. Zhenzhou: Elephant Press. (in Chinese)

HPIACR (Henan Provincial Institute of Archaeology and Cultural Relics) and MCH (Museum of Chinese History). 1992. *Dengfeng Wangchenggang yu Yangcheng (Excavation at Wangchenggang and Yangcheng)*. Beijing: Wenwu Press. (in Chinese)

HPIACR (Henan Provincial Institute of Archaeology and Cultural Relics) and PMACR (Puyang Municipal Institute of Archaeology and Cultural Relics). 2017. *Puyang Xishuipo (Excavation at Xishuipo site, Puyang)*. Henan: Zhongzhou Guji Press. (in Chinese)

Huang, Y. 2003. "Miaozigou yu Dabagou yizhi dongwu yihai jianding baogao (Faunal analysis of Miaozigou and Dabagou site)." *Miaozigou yu Dabagou (Excavation at Miaozigou and Dabagou site)*, edited by Inner Mongolia Provincial Institute of Archaeology and Cultural Relics, 599–611. Beijing: Zhongguo Dabaikequanshu Press. (in Chinese)

Hunn, E. 1975. "A measure of the degree of correspondence of folk to scientific biological classification." *American Ethnologist* 2 (2): 309–27.

— 2007. "Ethnobiology in four phases." *Journal of Ethnobiology* 27 (1): 1–10.

Ingold, T. 1980. *Hunters, Pastoralists and Ranchers : Reindeer Economies and Their Transformations*. Cambridge: Cambridge University Press.

Institute of Archaeology, CASS (Chinese Academy of Social Science). 1958. "Luoyang Donggangou yizhi fajue jianbao (A brief excavation report of the Donggangou site)." *Kaogu* (10): 537–40. (in Chinese)

— 1999. *Yanshi Erlitou 1959 Nian – 1978 Nian Kaogu Fajue Baogao (Archaeology Report on the Excavations of the Years 1959-1978 at Erlitou, Yanshi)*. Beijing: Zhongguo Dabaikequanshu Press. (in Chinese)

— 2003. *Zhongguo Kaoguxue. Xiashangzhou Juan (Chinese Archaeology. Xia, Shang and Zhou vol.)*. Beijing: Chinese Social Science Press. (in Chinese)

— 2007. *Yinxu de faxian yu yanjiu* (Discoveries and research on Yinxu). Beijing: Fangzhi Press. (in Chinese)

— 2010. *Zhongguo Kaoguxue. Xinshiqi Shidai Juan (Chinese Archaeology. Neolithic vol.)*. Beijing: Chinese Social Science Press. (in Chinese)

— 2017. *Zhishan Guanjunmiao (Excavation of the Guanjunmiao site)*. Beijing: Wenwu Press. (in Chinese)

Institute of Archaeology, CASS (Chinese Academy of Social Science) MCH (Museum of Chinese History) and SPIACR (Shanxi Provincial Institute of Archaeology and Cultural Relics). 1988. *Xiaxia Dongxiaping (Excavation of the Dongxiaping site)*. Beijing: Wenwu Press. (in Chinese)

Institute of Archaeology, CASS (Chinese Academy of Social Science) and ZMIACR (Zhengzhou Municipal Institute of Archaeology and Cultural Relics). 2009. "Henan Xinmi shi Xinzhai yizhi qianxueshi daxing jianzhu jizhi de fajue (Excavation of the shallow-pit-like structure at the Xinzhai site, Xinmi, Henan Province)." *Kaogu* 2: 32–47. (in Chinese)

Institute of Archaeology, CASS (Chinese Academy of Social Science) and LMBC (Linfen Municipal Bureau of Cultural Relics). 2015. *Xiangfen Taosi 1978-1985 (Excavation at Taosi 1978-1985)*. Beijing: Wenwu Press. (in Chinese)

Jia, Z., Y. Kuang and T. Jiao. 1983. "Yuxian Wadian yizhi fajue jianbao (A brief excavation report of the Wadian site)." *Wenwu* (3): 37–48. (in Chinese)

Jiang, H., X. Li, Y. Zhao, D. Ferguson, F. Hueber, S. Bera, Y. Wang, L. Zhao and C. Liu. 2006. "A new insight into Cannabis sativa (Cannabaceae) utilization from 2500-year-old Yanghai Tombs, Xinjiang, China." *Journal of Ethnopharmacology* 108: 414–22.

Jing, Y. and R. Flad. 2002. "Pig domestication in ancient China." *Antiquity* 76 (293): 724–32.

— 2006. "Research on early horse domestication in China." In *Equids in Time and Space: Papers in Honor of Vera Eisenmann*, edited by M. Mashkour, 124–31.

Jing, Y., R. Blench. and J. Huang. 2008. "Livestock in Ancient China: an archaeozoological perspective." In *Past Human Migrations in East Asia: Match Archaeology, Linguistics and Genetics*, edited by A. Sanchez-Mazas, R. Blench, M. D. Ross, I. Peiros and M. Lin, 427–44. London: Taylor and Francis.

Jones, M. K., H. Hunt, E. Lightfoot, D. Lister. and X. Liu. 2011. "Food globalization in prehistory" *World Archaeology* 43 (4): 665–75.

Jones, M.K. 2002. "Bio-archaeology and the Proto-Indo-European lexicon: the Kurgan hypothesis revisited." In *Ancient Interactions: East and West in Eurasia*, edited by K. Boyle, C. Renfrew, and M. Levine, 293–97. Cambridge: McDonald Institute for Archaeological Research.

Kamp, K. 1998. "Social hierarchy and burial treatments: a comparative assessment." *Cross-Cultural Research* 32 (1): 79–115.

— 2001. "Where have all the children gone?: the archaeology of childhood" *Journal of Archaeological Method and Theory* 8 (1): 1–34.

Keightley, D. N. 1978. *Sources of Shang History: the Oracle Bone Inscriptions of Bronze Age China*. Berkeley: University of California Press.

— 2000. *The Ancestral Landscape: Time, Space, and Community in Late Shang China*. Berkeley: IEAS & CCS.

Koerner, L. 1996. "Carl Linnaeus in his time and place." In *Cultures of Natural History*, edited by N. Jardine, J. Secord, and E. Spary, 145–62. Cambridge: Cambridge University Press.

— 1999. *Linnaeus: Nature and Nation*. Cambridge Mass: Harvard University Press.

Kesner, L. 1991. "The Taotie reconsidered: meanings and functions of the Shang theriomorphic imagery." *Artibus Asiae* 51: 29–53.

Lai, S.-J., Y.-P. Liu., Y.-X. Liu., X.-W. Li. and Y.-G. Yao. 2006. "Genetic diversity and origin of Chinese cattle revealed by mtDNA D-loop sequence variation." *Molecular Phylogenetics and Evolution* 38 (1): 146–54.

Larson, G., K. Dobney, U. Albarella and M. Fang. 2005. "Worldwide phylogeography of wild boar reveals multiple centers of pig domestication." *Science* 307 (5715): 1618–21.

Larson, G., E. Karlsson and A. Perri. 2012. "Rethinking dog domestication by integrating genetics, archeology, and biogeography." *Proceedings of the National Academy of Sciences* 109 (23): 8878–83.

Larson, G., R. Liu, X. Zhao. and J. Yuan. 2010. "Patterns of East Asian pig domestication, migration, and turnover revealed by modern and ancient DNA." *Proceedings of the National Academy of Science* 107 (17): 7686–91.

Lee-Thorp, J. 2008. "On isotopes and old bones." *Archaeometry* 50 (2): 925–50.

Lee, G.-A., G. W. Crawford., L. Liu. and X. Chen. 2007. "Plants and people from the Early Neolithic to Shang periods in North China." *Proceedings of the National Academy of Sciences* 104 (3). National Academy of Sciences: 1087–92.

Lee, Y.K., S. Allan, F. Allard, J. An, G. Ban, S. Bowman, C.E. Buck, J.B. Kenworthy, C.D. Litton, A.F.M. Smith, Y. Cai, L. Fu, K. Chang, C.C. Lamberg-Karlovsky, C. Chen, Z. Xi, C.L.N. Ruggles, N.J. Saunders, L. Liu, C. Xu, *et al.* 2002. "Building the chronology of early Chinese history." *Asian Perspectives* 41 (1): 15–42.

Legge, A. 1885. *The LÎ KÎ (The Book of Rites Part II, Sacred Books of the East vol. 27)*. Retrieved on 12-12-2016 from Internet Sacred Text Archive (ISTA. www.sacred-texts.com/cfu/liki2/liki220.htm%0A.

Lehmann, W.P. 1962. *Historical Linguistics : An Introduction*. London: Holt, Rinehart and Winston.

Lei, C., R. Su, M. Bower and C. Edwards. 2009. "Multiple maternal origins of native modern and ancient horse populations in China." *Animal Genetics* 40 (6): 933–44.

Li, B. 2009. "Zhongguo gudai wenming yanjin de liangzhong moshi – Hongshan, Liangzhu, Yangshao damu yuqi guancha suxiang (Two modes of civilisation formation in ancient China – comparing grave jades from Hongshan, Liangzhu and Yangshao culture)." *Wenwu* (3):47-56. (in Chinese)

Li, C. and Y. Liao. 1982. "Zhenzhou Mazhuang Longshan wenhua yizhi fajue jianbao (A brief excavation of the Longshan period Mazhuang site)." *Zhongyuan Wenwu* (4): 24–31. (in Chinese)

Li, F. 2014. "*Yuxi yizhi T0304 tanfang dongwu gugue de chubu guancha yu fenxi* (The observation and analysis of animal fossils from Yusi site about T0403 Excavation squares)." Master diss., Chongqing Normal University.

Li, H. 2010. "Zhongguo gudian yiyi de datong, xiaokang shehui (Datong and Xiaokang society in ancient China)." *Shehuixue Yanjiu* (4): 126–42. (in Chinese)

Li, L. 2013. "Xinzhai chengzhi de juluo xingzhi tanxi (Exploring the nature the Xinzhai walled settlement)." *Zhongzhou Xuekan* (6): 118–21. (in Chinese)

Li, W. 2002. "Erlitou wenhua yi qi yicun yu Xia wenhua chushi (Phase 1 of the Erlitou culture and the inception of the Xia culture)." *Zhongyuan Wenwu* (1): 33–42. (in Chinese)

Li, X. 2011. "Shang wenhua muzang zhong suizang de gou sheng yanjiu er ti (Studies on dog sacrifice found in Shang tombs)." *Nanfang Wenwu* (2): 100–4. (in Chinese)

Li, X. 2002. "The Xia-Shang-Zhou Chronology Project: methodology and results." *Journal of East Asian Archaeology* 4 (1). Brill: 321–33.

Li, Y., C. Zhang, W. Taylor, T. Timothy, L. Chen, R. Flad, N. Boivin, H. Liu, Y. You, J. Wang *et al.* 2020. "Early evidence for mounted horseback riding in northwest China." *Proceedings of the National Academy of Sciences* 117: 29569.

Li, Z., R. Campbell, K. Brunson, J. Yuan and Y. Tao. 2014. "The exploitation of domestic animal products from the Late Neolithic Age to the Early Bronze Age in the heartland of ancient China." In *Animal Secondary Products: Domestic Animal Exploitation in Prehistoric Europe, the Near East and the Far East*, edited by H. Greenfield, 56–79. Oxford: Oxbow Books.

Li, Z. and R. Campbell. 2019. "Puppies for the ancestors: the many roles of Shang dogs." *Archaeological research in Asia* 17: 161–72.

Linares, O. 1976. "'Garden hunting' in the American tropics." *Human Ecology* 4 (4): 331–49.

Lindblad-Toh, K., C.M. Wade, T.S. Mikkelsen, E.K. Karlsson, D.B. Jaffe, M. Kamal, M. Clamp, J.L. Chang, E.J. Kulbokas, M.C. Zody, E. Mauceli, X. Xie, M. Breen, R.K. Wayne, E.A. Ostrander, C.P. Ponting, F. Galibert, D.R. Smith, P.J. DeJong, E. Kirkness, *et al.* 2005. "Genome sequence, comparative analysis and haplotype structure of the domestic dog." *Nature* 438 (7069): 803–19.

Linduff, K. 2003. "A walk on the wide side: Late Shang appropriation of horse in China." In *Prehistoric Steppe Adaptation and the Horse,* edited by M. Levine, C. Renfrew, and K.V. Boyle, 139–62. Oxford: Oxbow Books.

Liu, C. 2016. "Review on the studies of unearthed Mawangdui medical books." *Chinese Studies* 5 (1): 6–14.

Liu, L. 1996. "Mortuary ritual and social hierarchy in the Longshan Culture." *Early China* 21: 1–46.

—2004. *The Chinese Neolithic: Trajectories to Early States*. Cambridge: Cambridge University Press.

—2009. "State emergence in early China." *Annual Review of Anthropology* 38 (1): 217–32.

Liu, L. and X. Chen. 2001. Settlement archaeology and the study of social complexity in China, *Review of Archaeology* 22 (2): 4–22.

—2003. *State Formation in Early China*. London: Duckworth.

—2006. Sociopolitical change from Neolithic to Bronze Age China, in M. Stark (*ed.*), *Archaeology of Asia*: 149–76. Oxford, UK: Blackwell Publishing Ltd.

—2012. *The Archaeology of China: from the Late Paleolithic to the early Bronze Age*. Cambridge: Cambridge University Press.

Liu, L. and H. Xu. 2007. "Rethinking Erlitou: legend, history and Chinese archaeology." *Antiquity* 81 (314): 886–901.

Liu, L., Y. Yan and X. Qin. 2001. "Shaanxi Lintong Kangjia Longshan wenhua yizhi 1990nian fajue dongwu yicun (Faunal report of the 1991 excavation at the Kangjia site)." *Huaxia Kaogu* (1): 3–24. (in Chinese)

Liu, X. and M.K. Jones. 2014. "Under one roof: People, crops and animals in Neolithic North China." In *Living in the Landscape: Essays in Honour of Graeme Barker*, edited by K. Boyle, R.J. Rabett, and C.O. Hunt, 227–34. Cambridge: McDonald Institute Monographs.

Liu, Y. 2014. "Shangzhou jisi dongwu yanjiu zongshu (A review on sacrificial animals used during the Shang and Zhou Dynasties)." *Nanfang Wenwu* (1): 58–64. (in Chinese)

Luo, Y. 2012. *Zhongguo gudai zhulei xunhua yu yishixing shiyong (The Domestication, Raising, and Ritual Use of Pig in Ancient China)*. Beijing: Kexue Press. (in Chinese)

Luo, Y. and J. Zhang. 2008. "Henan Wuyang Jiahu yizhi chutu zhugu de zai yanjiu (Reanalysis of pig remains from the Jiahu site)." *Kaogu* (1): 90–6. (in Chinese)

Lv, P. 2007. "Lun Zhongguo gudai jiayang huangniu de qiyuan (On the Origin of Cattle in Ancient China)." Master diss., Institute of Archaeology, CASS.

—2009. "Yuzhou wadian yizhi dongwu guge de jianding he yanjiu (Zooarchaeological study on faunal assemblage from the Wadian site)." In *Zhonghua Wenming Tanyuan Gongcheng Wenji – Jishu yu Jingji Juan (Discover the beginning of Chinese Civilization – the Volume of Technology and Economy)*, edited by Department of Social Development and Department of Museums and Cultural Relics, Bureau of Cultural Relics, China, 179–94. Beijing: Kexue Press. (in Chinese)

Lyman, R. 1994. *Vertebrate taphonomy.* Cambridge: Cambridge University Press.

—2008. *Quantitative Paleozoology*. Cambridge: Cambridge University Press.

Ma, Q. and G. Cao. 1983. "Henan Huaiyang Pingliangtai Longshan wenhua chengzhi shijue baogao (Excavation of the Longshan site Pingliangtai in Henan Province)." *Wenwu* (3): 21–36. (in Chinese)

Ma, S. 2004. "Xinmi Xinzhai chengzhi yu Xia Qi zhi ju (The Xinzhai site and the Qi captial of the Xia dynasty)." *Zhongyuan Wenwu* (3): 12–6. (in Chinese)

—2007. "Xinzhai chengzhi yu Qi du Xia yi wenti tansuo (Exploring the relationship between the Xinzhai site and the Qi captial and Xia cities)." *Kaogu yu Wenwu* (3): 54–8. (in Chinese)

—2008. "Dengfeng Wangchenggang chengzhi yu yudu Yangcheng (Wangchenggang walled site and Emperor Yu's Capital Yang city)." *Zhongyuan Wenwu* (2): 22–6. (in Chinese)

Ma, X. 2010. "Guanyu Zhongguo guqi yanjiu de gi ge wenti (Discussion of a few questions related to Chinese bone artifact research)." *Huaxia Kaogu* (2): 138–42. (in Chinese)

Manketlow, M. n.d. History of taxonomy. http://atbi.eu/summerschool/files/summerschool/Manktelow_Syllabus.pdf

Marciniak, A. 2011. "Folk taxonomies and human-animal relations: the early Neolithic in the Polish Lowlands." In *Ethnozooarchaeology: the Present and Past of Human-Animal Relationships*, edited by U. Albarella and A. Trentacoste, 29–40. Oxford: Oxbow Books.

Mayr, E. 1942. *Systematics and the Origin of Species, from the Viewpoint of a Zoologist*. Cambridge Mass: Harvard University Press.

Mayr, E. and W. J. Bock. 2002. "Classifications and other ordering systems." *Journal of Zoological Systematics and Evolutionary Research* 40 (4): 169–94.

MCH (Museum of Chinese History). 1996. *Yuanqu Shangcheng—1985-1986niandu Kancha Baogao (Yuanqu Shangcheng—Excavation during 1985-1986)*. Beijing: Kexue Press. (in Chinese)

Medin, D.L. and S. Atran. 1999. *Folkbiology*. Cambridge, Massachusetts: The MIT Press.

Miller, N. 2013. "Agropastoralism and archaeobiology: connecting plants, animals and people in west and central Asia." *Environmental Archaeology* 18 (3): 247-56

Miracle, P. and L. Pugsley. 2006. "Vertebrate faunal remains from Pupićina Cave." In *Prehistoric Herders of northern Istria: the Archaeology of Pupićina Cave. Volume 1*, edited by P.T. Miracle and S. Forenbaher, 259–399. Pula: Archeoloski Muzej Istre.

Mo, L. 2016. "Changjiang zhongxiayou diqu shiqian yulei yicun chubu yanjiu (The preliminary analysis of prehistoric fish remains from the Middle and Lower Yangtze Valley)." *Nanfang Wenwu* (4): 223–33. (in Chinese)

Moore, H. L. 1982. "The interpretation of spatial patterning in settlement residues, in *Symbolic and Structural Archaeology*: 74–79. Cambridge: Cambridge University Press.

— 1986. *Space, Text and Gender: An Anthropological Study of the Marakwet of Kenya*. Cambridge: Cambridge University Press.

Morris, B. 1998. *The Power of Animals : An Ethnography*. Oxford, UK: Berghahm Books.

— 2000. "The pragmatics of folk classification." In *Ethnobotany: the reader*, edited by P. Minnis, 69–87. Norman: University of Oklahoma Press.

Morris, J. 2010. "Associated bone groups: beyond the Iron Age." In *Integrating Social and Environmental Archaeologies: Reconsidering Deposition*, edited by J. Morris and M. Matby, 12–23. BAR Publishing International Series 2077.

National Cultural Heritage Administration Three Georges Archaeological Team. 2001. *Chaotianzui yu Zhongbaodao (The Chaotianzui and Zhongbaodao sites)*. Beijing: Wenwu Press. (in Chinese)

Netting, R., R. Wilk and E. Arnould. 1984. *Households: Comparative and Historical Studies of the Domestic Group*. Berkeley: University of California Press.

Neusius, S. 2008. "Game procurement among temperate horticulturists: the case for garden hunting by the Dolores Anasazi." In *Case Studies in Environmental Archaeology*, edited by E. Reitz, C. Scarry, and S. Scudder. New York: Springer.

Noegel, S. 2019. "From ape to zebra: on wild animals and taxonomy in ancient Israel." In *Animals and their Relations to Gods, Humans and Things in the Ancient World*, edited by R. Mattila, S. Ito and S. Fink, 95–136. Wiesbaden: Springer VS.

O'Connor, T.P. 1988. *Bones from the General Accident Site, Tanner Row, the Archaeology of York (vol.15)*. London: Council for British Archaeology.

— 1997. "Working at relationships: another look at animal domestication." *Antiquity* 71 (271): 149–56.

— 2003. *The Analysis of Urban Animal Bone Assemblages: a Handbook for Archaeologists*. York: Council for British Archaeology.

— 2013. "Humans and animals: refuting Aquinas." *Archaeological Review from Cambridge* 28 (2): 186–94.

Okamura, H. 2008. *Chūgoku Bunmei: Nōgyō to Rei-sei no Kōkogaku (Chinese Civilisation: Archaeology of Agriculture and Ritual)*. Kyoto: Kyoto University Press. (in Japanese)

Olsen, S.L. 2006. "Early horse domestication on the Eurasian steppe." In *Documenting Domestication: New Genetic and Paradigms,* edited by S.L. Olsen, S. Grant, A.M. Choyke, and L. Bartosiewicz, 81–113. Oxford: BAR International Series 1560.

Olsen, S.J. 1988. "The horse in ancient China and its cultural influence in some other areas." *Proceedings of the Academy of Natural Sciences of Philadelphia* 140: 151–89.

Outram, A. 2001. "A new approach to identifying bone marrow and grease exploitation: why the 'indeterminate' fragments should not be ignored." *Journal of Archaeological Science* 28 (4): 401–10.

Pang, J., C. Kluetsch, X. Zou, A. Zhang and L. Luo. 2009. "MtDNA data indicate a single origin for dogs south of Yangtze River, less than 16,300 years ago, from numerous wolves." *Molecular Biology and Evolution* 26 (12): 2849–64.

Pang, X. and J. Gao. 2008. "Guanyu Xinzhai qi yicun yanjiu de jige wenti (Issues about the Xinzhai phase)." *Huaxia Kaogu* (1): 73–80. (in Chinese)

Partridge, E. 2002. *Origins: A Short Etymological Dictionary of Modern English*. Oxon: Routledge.

Pearce, M. 2008. "Structured deposition in early Neolithic northern Italy." *Journal of Mediterranean Archaeology* 21 (1): 19–33.

Pechenkina, E., S. Ambrose and X. Ma. 2005. "Reconstructing northern Chinese Neolithic subsistence practices by isotopic analysis." *Journal of Archaeological Science* 32 (8): 1176–89.

Pei, M. 1993. "Zhenzhou Shangdai taozifu shishi (Study on the pottery marks of the Shang Dynasty in Zhenzhou region)." In *Heluo Wenming Lunwen Ji (Essays on the Heluo Civilization)*, edited by 2nd Luoyang Municipal Archaeological Team (*ed.*), Zhenzhou: Zhongzhou Guji Press. (in Chinese)

Peled, I. 2019. "Categorization and hierarchy: animals and their relations to gods, humans and things in the Hittite World." In *Animals and their Relations to Gods, Humans and Things in the Ancient World,* edited by R.

Mattila, S. Ito and S. Fink: 79–94. Wiesbaden: Springer VS.

Peterson, C. and G. Shelach. 2012. "Jiangzhai: Social and economic organization of a Middle Neolithic Chinese village." *Journal of Anthropological Archaeology* 31 (3): 265–301.

Pickering, T., C. Marean and M. Dominguez-Rodrigo. 2003. "Importance of limb bone shaft fragments in zooarchaeology: a response to 'On in situ attrition and vertebrate body part profiles'(2002), by MC Stiner." *Journal of Archaeological Science* 30 (11): 1469–82.

Piggott, S. 1974. "Chariots in the Caucasus and in China." *Antiquity* 48 (18): 16–24.

Pike, K. 1967. *Language in Relation to a Unified Theory of the Structure of Human Behavior (2nd revision)*. The Hague, Netherlands: Mouton & Co.

Pollard, J. 1995. "Inscribing space: formal deposition at the later Neolithic monument of Woodhenge, Wiltshire." *Proceedings of the Prehistoric Society* 21: 137–56.

— 2001. "The aesthetics of depositional practice." *World Archaeology* 33: 315–33.

— 2006. "A Community of Beings: Animals and People in the Neolithic Southern Britain." In *Animals in the Neolithic of Britain and Europe*, edited by D. Serjeantson and D. Field, 135–48. Oxford: Oxbow Books.

Pollard, J. and C. Ruggles. 2001. "Shifting perceptions: spatial order, cosmology, and patterns of deposition at Stonehenge." *Cambridge Archaeological Journal* 11 (1): 69-9.

Postgate, J.N. 2015. *Early Mesopotamia: Society and Economy at the Dawn of History*. New York: Routledge.

Qi, G. 1988. "Jiangzhai xinshiqishidai yizhi dongwuqun de fenxi (Faunal analysis of the Jiangzhai site)." In *Jiangzhai (Excavation Report of Jiangzhai site)*, edited by Xi'an Banpo Museum, Shaanxi Provincial Institute of Archaeology, and Lingtong County Museum, 504–39. Beijing: Wenwu Press. (in Chinese)

Qi, G., Z. Lin and J. An. 2006. "Dadiwan yizhi dongwu yicun jianding baogao (Faunal analysis of the Dadiwan site)." In *Qin'an Dadiwan (Excavation of the Dadiwan site)*, edited by APIACR (Anhui Provincial Institute of Archaeology and Cultural Relics), 861–910. Beijing: Wenwu Press. (in Chinese)

Queiroz, K. De. 1988. "Systematics and the Darwinian revolution." *Philosophy of Science* 55 (2): 238–59.

— 1997. "The Linnaean hierarchy and the evolutionization of taxonomy, with emphasis on the problem of nomenclature." *Aliso* 15 (2): 125–44.

Reinhart, K. 2015. "Religion, violence, and emotion: modes of religiosity in the Neolithic and Bronze Age of northern China." *Journal of World Prehistory* 28 (2): 113–77.

Renfrew, C. and B. Liu. 2018. "The emergence of complex society in China: the case of Liangzhu." *Antiquity* 92 (364): 975-90.

Richards, C. and J. Thomas. 1984. "Ritual activity and structured deposition in Later Neolithic Wessex." In *Neolithic Studies: A Review of Some Current Research*, edited by R. Bradley and J. Gardiner, 189–218. Oxford: BAR Publishing British Series 133.

Rong, G. 1985. *Jinwen Bian (Compilation of Oracle Bone Characters)*. Beijing: Zhonghua Shuju Press. (in Chinese)

Ross, B. and G. Murphy. 1999. "Food for thought: Cross-classification and category organization in a complex real-world domain." *Cognitive Psychology* 38 (4): 495–553.

Russell, N. 2002. "The wild side of animal domestication." *Society & Animals* 10 (3): 285–302.

— 2007. "The domestication of anthropology." In *Where the Wild Things are Now: Domestication Reconsidered*, edited by R. Cassidy and M. Mullin, 27–48. Oxford: BERG.

Russell, N. and B. During. 2006. "Worthy is the lamb: a double burial at Neolithic Çatalhöyük (Turkey)." *Paléorient* 3 (2): 73–84.

Russo, E., H. Jiang, X. Li, A. Sutton, A. Carboni, F. del Bianco, G. Manndolino, D. Potter, Y. Zhao, S. Bera *et al.* 2008. "Phytochemical and genetic analyses of ancient cannabis from Central Asia." *Journal of Experimental Botany* 59: 4171–82.

SAMPU (School of Archaeology and Museology in Peking University) and HPIACR (Henan Provincial Institute of Cultural Relics and Archaeology). 2007. *Dengfeng Wangchenggang (Excavation of Wangchenggang)*. Beijing: Elephant Press. (in Chinese)

Sanchez, C. S. 1993. "Funerary practices and human sacrifice in Teotihuacan burials." In *Teotihuacan: Art from the City of the Gods*, edited by K. Berrin and E. Pasztory, 109–15. New York: Thames and Hudson.

Savolainen, P., Y. Zhang, J. Luo, J. Lundeberg and T. Leitner. 2002. "Genetic evidence for an East Asia origin of domestic Dogs." *Science* 298 (5598): 1610-1613.

Schiebinger, L. 2003. "The philosopher's beard: women and gender in science." In *The Cambridge History of Science: Volume 4, Eighteenth-Century*, edited by R. Porter, 184–210. Cambridge: Cambridge University Press.

Schuessler, A. 2007. *ABC Etymological Dictionary of Old Chinese*. Honolulu: University of Hawaii Press.

Serjeantson, D. 2000. "Good to eat and good to think with: classifying animals from complex sites." In *Animal Bones, Human Societies*, edited by P. Rowley-Conwy, 179–88. Oxford: Oxbow Books.

Serpell, J. 1995. *The Domestic Dog: its Evolution, Behaviour and Interactions with People.* Cambridge: Cambridge University Press

Service. E. R. 1975. *Origins of the State and Civilization: the Process of Cultural Evolution.* New York: Norton.

Shaughnessy, E. L. 2009. "Chronologies of ancient China: A critique of the 'Xia-Shang-Zhou chronology project," In *Windows on the Chinese World: Reflections by Five Historians*, edited by W. Ho, 15–28. Lanham: Lexington Books.

Shelach, G. 2006. "Economic adaptation, community structure, and sharing strategies of households at early sedentary communities in northeast China." *Journal of Anthropological Archaeology* 25 (3): 318–45.

Shelach, G. and Y. Jaffe. 2014. "The earliest states in China: a long-term trajectory approach." *Journal of Archaeological Research* 22 (4): 327–64.

Sheng, H., K. Cao, W. Li, Y. Ma, H. Xu and E. Zhang. 1992. *Zhongguo Lulei Dongwu (Deer in China).* Shanghai: Huadong Shifan Daxue Press. (in Chinese)

Sherratt, A. 1991. "Review of Ian Hodder 'The Domestication of Europe'." *Antiquity* 65 (258): 742–43.

Shi, Z. 1976. *Xiaotun yizhi de faxian yu fajue: bing bian, Yinxu muzang zhi si, yiqu jizhi shangxia de1 muzang (Xiaotun: The Yin-Shang site at Anyang, Henan. Vol 1: The site, its discovery and excavations. Fascicle 3: Burials above and underneath the B area Foundations.* Taipei: Institute of History and Philology, Academia Sinca. (in Chinese)

Shih, C. 1953. "Henan Anyang Xiaotun mu zhong de dongwu yihai (Animal remains in tombs at the Xiaotun Site)."*Wenshizhe Xuebao* (5): 1–14. (in Chinese)

Shihe Kaogu Dui (Shihe Archaeological Team). 1990. "Hubeisheng shihe yizhiqun 1987nian fajue jianbao (The excavation report of the Shehe culture sites in 1987)." *Wenwu* (8): 6–14. (in Chinese)

Shotwell, J. 1958. "Inter-community relationships in Hemphillian (Mid-Pliocene) mammals." *Ecology* 39 (2): 271–82.

Sillar, B. 1994. "Playing with god: cultural perceptions of children, play and miniatures in the Andes." *Archaeological Review from Cambridge* 13 (2): 47–63.

Skoglund, P., E. Ersmark, E. Palkopoulou, L. Dalén, C. Vilà, P. Savolainen, J.E. Maldonado, I.R. Amorim, J.E. Rice, R.L. Honeycutt, K.A. Crandall, J. Lundeberg, R.K. Wayne, C. Vila, I.R. Amorim, J.A. Leonard, D. Posada, J. Castroviejo, F. Petrucci-Fonseca, K.A. Crandall, *et al.* 2015. "Ancient wolf genome reveals an early divergence of domestic dog ancestors and admixture into high-latitude breeds." *Current Biology* 25 (11): 1515–9.

Stearn, W. 1959. "The background of Linnaeus's contributions to the nomenclature and methods of systematic biology." *Systematic Zoology* 8 (1): 4–22.

— 1973. *Botanical Latin: History, Grammar, Syntax, Terminology and Vocabulary Botanical Latin.* Newton Abbot: David and Charles.

Sterckx, R. 1996. "An ancient Chinese horse ritual." *Early China* (21): 47–79.

— 2002. *The Animal and the Daemon in Early China.* New York: State University of New York Press.

— 2005. "Animal classification in ancient China." *East Asian Science, Technology, and Medicine* (23): 26–53.

Stiner, M. 1991. "Food procurement and transport by human and non-human predators." *Journal of Archaeological Science* 18 (4): 455–82.

Storey, G. R. 2006. *Urbanism in the Preindustrial World : Cross-Cultural Approaches.* Tuscaloosa: University of Alabama Press.

Sun, D., Y. Liu and G. Chen. 1981. "Hebei Wu'an Cishan yizhi (Excavation of the Cishan site, Hebei Province)." *Kaogu Xuebao* (3): 303–38. (in Chinese)

Sun, H. 1965. *Jiaguwen Bian (Compilation of Oracle Bone Characters).* Beijing: Zhonghua Shuju Press. (in Chhinese)

Sykes, N. 2014. *Beastly Questions: Animal Answers to Archaeological Issues.* London: Bloomsbury Publishing.

Takatsuki, S., K. Suzuki and H. Higashi. 2000. "Seasonal elevational movements of sika deer on Mt. Goyo, northern Japan." *Mammal Study* 25 (20): 107–14.

Thomas, J. 1999. *Understanding the Neolithic.* London: Routledge.

— 2012. "Some deposits are more structured than others." *Archaeological Dialogues* 19 (2): 124–27.

Thomas, K. 1983. *Man and the Natural World: Changing Attitudes in England, 1500-1800. Allen Lane.* London.

Thorp, R. L. 2006. *China in the Early Bronze Age : Shang Civilization.* Philadelphia: University of Pennsylvania Press.

Trigger. B. G. 1999. "Shang political organization: A comparative approach." *Journal of East Asian Archaeology* (1): 43–62.

Tu, Y., Y. Jiang, Z. Han, W. Feng, S. Deng, Y. Wang, J. Li, L. Lu, Y. Zhang, Y. Zhu, B. Han, B. Lu, X. Pan and H. Wei. 1989. *Zhongguo yang pinzhong zhi (Sheep and goat breeds in China).* Shanghai: Scientific & Technical Publishers. (in Chinese)

Underhill, A. 1989. "Warfare during the Chinese Neolithic period: a review of the evidence." In *Cultures in Conflict: Current Archaeological Perspectives : Proceedings of the 20th Annual Conference of the*

Archaeological Association of the University of Calgary, edited by D.C. Tkaczuk and B.C. Vivian, 229–37. Alberta: Archaeological Association of the University of Calgary.

— 1991. "Pottery production in chiefdoms: The Longshan period in northern China." *World Archaeology* 23 (1): 12–27.

Vila, C., P. Savolainen, J.E. Maldonado, I.R. Amorim, J.E. Rice, R.L. Honeycutt, K.A. Crandall, J. Lundeberg and R.K. Wayne. 1997. "Multiple and ancient origins of the domestic dog." *Science* 276 (5319): 1687–9.

Von Holdt, B.M., J.P. Pollinger, K.E. Lohmueller, E. Han, H.G. Parker, P. Quignon, J.D. Degenhardt, A.R. Boyko, D.A. Earl, A. Auton, *et al.* 2010. "Genome-wide SNP and haplotype analyses reveal a rich history underlying dog domestication." *Nature* 464 (7290): 898–902.

Wang, B. 2006. *Jiaguwen Xiao Zidian (Little Dictionary of Oracle Bone Script)*. Beijing: Wenwu Press. (in Chinese)

Wang, H., W. Wang and S. Hu. 2014. "Yangshao shidai renlei shoulie meihualu de celue: yi Tongchuan Wayaogou yizhi wei anli (Deer hunting during the Yangshao period: a case study of the Wayaogou site)." *Acta Anthropologica Sinica* 33 (1): 90–100. (in Chinese)

Wang, J. 2005. "Shiqian shiqi de diaoke yu taosu (On the prehistoric carvings and figurines)." *Dongnan Wenhua* (1): 1–5. (in Chinese)

Wang, K. 1982. "Shilun woguo renji he renxun de qiyuan (On the origin of human sacrifice in China)." *Wenwu* (2): 69–72. (in Chinese)

Wang, L. 1999. "Shi lun Longshan wenhua shidai de renxun he renji (On human sacrifices during the Lognshan period)." *Dongnan Wenhua* (4): 22–7. (in Chinese)

Wang, T. 2007. "Shang ritual animals: colour and meaning (part 1)." *Bulletin of the School of Oriental and African Studies* 70 (2): 305–72.

Wang, Y., S. Wang, S. Li, H. Chen and T. Luo. 1966. "Henan Zhenzhou Shangjie Shangdai yizhi fajue baogao (Excavation of the Shangjie site)." *Kaogu* (1): 1–7. (in Chinese)

Wapnish, P. 1995. "Towards establishing a conceptual basis for animal categories in archaeology." In *Methods in the Mediterranean: Historical and Archaeological Views on Text and Archaeology*, edited by D. Small, 233–73. Leiden: The Netherlands.

Watson, J. 1979. "The estimation of the relative frequencies of mammalian species: Khirokitia 1972." *Journal of Archaeological Science* 6 (2): 127–37.

Wedel. W. R. 1938. *The Direct-Historical Approach in Pawnee archeology: (with six plates)*. Washington DC: Smithsonian Institution.

Wei, X. 2008. "Puchengdian Erlitou chengzhi ruangan wenti taolun (Issues about the Puchengdian site)." *Zhongyuan Wenwu* (3): 36–43. (in Chinese)

White, C., M. Pohl and H. Schwarcz. 2001. "Isotopic evidence for Maya patterns of deer and dog use at Preclassic Colha." *Journal of Archaeological Science* 28 (1): 89–107.

White, T. 1953. "A method of calculating the dietary percentage of various food animals utilized by aboriginal peoples." *American Antiquity* 18 (4): 396–98.

Whitehouse, H. 1995. *Inside the Cult : Religious Innovation and Transmission in Papua New Guinea*. Oxford: Oxford University Press.

— 2002. "Modes of religiosity: towards a cognitive explanation of the sociopolitical dynamics of religion." *Method and Theory in the Study of Religion* 14 (3-4): 293–315.

— 2004. *Modes of Religiosity: A Cognitive Theory of Religious Transmission*. Oxford: Altamira Press.

Wilk, R. and W. Rathje. 1982. "Household archaeology." *The American Behavioral Scientist (pre-1986)* 25 (6): 617.

Wolf. E. R. 1982. *Europe and the People Without History*. Berkeley: University of California.

Worster, D. 1977. *Nature's Economy: the Roots of Ecology*. San Francisco: Sierra Club Books.

Wu, H. 1979. "Yi zu zaoqi de yushi diaoke (A group of early stone and jade objects)." *Meishu Yanjiu* 1: 64-70.

Xia-Shang-Zhou Chronology Project Team. 2001. *Xia Shang Zhou duandai gongcheng (Xia-Shang-Zhou Chronology Project)*. Beijing: Shijie Tushu Press. (in Chinese)

Xia, N. 1985. *Zhongguo Wenming de Qiyuan (The Origins of Chinese Civilization)*. Beijing: Wenwu Press. (in Chinese)

Xu, H. 2004. "Erlitou yizhi fajue he yanjiu de huigu yu sikao (Erlitou revisited)." *Kaogu* (11): 32–38. (in Chinese)

Xu, J., Y. Huan and M. Kong. 2016. "Wolong ziran baohuqu yesheng dongwu zhaoshi nongdi tezheng ji yingxiang jizhi (The characteristics of agricultural fields damaged by wildlifes)." *Shengtaixue Bao* 36 (12): 3748–57. (in Chinese)

Xu, S. 1963. *Shuowen Jiezi (Explaining Graphs and Analyzing Characters)*. Beijing: Zhonghua Shuju Press. (in Chinese)

Xu, X. 1959. "1959 nian xia yuxi diaocha 'Xia' xu de chubu baogao (Preliminary survey of the 'Xia' remnants in west Henan Province)." *Kaogu* (11): 592–600. (in Chinese)

Xu, X. 1989. *Yangshao Wenhua Yanjiu* (Study on the Yangshao Culture). Beijing: Wenwu Press. (in Chinese)

Yan, W. 1989. *Yangshao Wenhua Yanjiu (Study of the Yangshao Culture)*. Beijing: Wenwu Press. (in Chinese)

Yan, W. and Y. Li. 1961. "Luoyang Wangwan yizhi fajue jianbao (Excavation report of the Wangwan site)." *Kaogu* (11): 175–81. (in Chinese)

Yang, H. and J. Ding. 2000. "Wushan Daxi yizhi de kaogu faxian yu yanjiu(Archaeological discoveries and research at the Daxi site, Wushan)." *Sichuan Wenwu* (1): 9–19. (in Chinese).

Yang, Z. and D. Liu. 1949. "Anyang Yinxu burudongwu sun buyi (Faunal remains from the Yinxu site: a supplementary)." *Kaogu Xuebao* (4): 145–53. (in Chinese)

Yoffee, N. 2005. *Myths of the Archaic State: Evolution of the Earliest Cities, States, and Civilizations*. Cambridge: Cambridge University Press.

Young, C.C. and D. Liu. 1949. "Anyang Yinxu zhi burudongwuqun buyi (More on the Yinxu fauna from the Anyang site)." *Kaogu Xuebao* 4. (in Chinese)

Yuan, G. 1991. "Linru Meishan yizhi 1987-88nian fajue baogao (Excavation of the Mishap site during 1987-1988)." *Huaxia Kaogu* 3: 5–23. (in Chinese)

Yuan, G. 2000. "Mengzhuang Longshan wenhua yichun yanjiu (Study of the Mengzhuang site, Longshan period)." *Kaogu* (3): 21–38. (in Chinese)

Yuan, J. and N. Dong 2018. "Zhongguo jiayang dongwu qiyuan de zaisikao (Rethinking the origins of animal domestication in China)." *Kaogu* (9): 113–20. (in Chinese)

Yuan, J. and R. Flad. 2005. "New zooarchaeological evidence for changes in Shang Dynasty animal sacrifice." *Journal of Anthropological Archaeology* 24 (3): 250–77.

Yuan, J. 2010. "Zhongguo gudai jiayang dongwu de dongwukaoguxue yanjiu (Zooarchaeological study on the domestic animals in ancient China)." *Disiji Yanjiu* (2): 298–306. (in Chinese)

Yuan, J., Y. Huang and M. Yang. 2007. "Gongyuanqian 2500 nian -- Gongyuanqian 1500 nian zhongyuan diqu dongwukaoguxue yanjiu -- Taosi, Wangchenggang, Xinzhai he Erlitou yizhi weili (Zooarchaeological study of the Central Plain during 2500--1500 BC.: case studies of Taosi, Wangchenggang, Xinzha." In *Keji Kaogu (vol. 2) (Archaeological Science, vol. 2),* edited by Center of Archaeological Science, Institute of Archaeology CASS, 12–34. Beijing: Kexue Press. (in Chinese)

Yuan, J. and J. Li. 2010. "Henbei Xushui Nanzhuangtou yizhi chutu dongwu yicun yanjiu baogao (Faunal analysis of the Nanzhuangtou site, Henbei Province)." *Kaogu Xuebao* (3): 385–91. (in Chinese)

Zeder, M. 2006. "Reconciling rates of long bone fusion and tooth eruption and wear in sheep (Ovis) and goat (Capra)." In *Recent Advances in Ageing and Sexing Animal Bones*, edited by D. Ruscillo, 87–118. Oxford: Oxbow Books.

Zhang, C., L. Fan, Q. Sun and M. Yang. 2000. "Henan dengzhou baligang yizhi 1998 nian fajue jianbao (1998's Excavation report of the Baligang site, Dengzhou, Henan Province)." *Wenwu* (11): 23–31. (in Chinese)

Zhang, X., S. Chou, L. Cai, G. Bo, G. Wang and J. Zhong. 2007. "Xinzhai – Erlitou – Erligang wenhua kaogu niandai xulie de jianli yu wanshan (Improving the chronology of Xinzhai Erlitou and Erligang culture)." *Kaogu* 1: 74–89. (in Chinese)

Zhang, X. and C. Zhao. 2015. "Xinzhai yizhi chutu bufen dongwu gu de tan dan wending tongweisu fenxi (Isotope analysis of faunal remains from the Xinzhai site)." *Nanfang Wenwu* (4): 232–40. (in Chinese)

Zhao, C. and Y. Fang. 2014. "Yuzhou wadian yizhi chutu bufen renlei yayouzhi de sitongweisu bizhi fenxi (The analysis of the Sr isotope results from human teeth at Wadian)." *Huaxia Kaogu* (3): 123–27. (in Chinese)

Zhao, C. 2001. *Zhenluo Diqu Xinshiqi shidai Juluo de Yanbian (Settlemtn Change in the Neolithic Zhenluo Region)*. Beijing: Peking University Press.

— 2004. "Xinmi Xinzhai yu Xia Qi zhi ju (The Xinzhai site and the Qi captial of the Xia Dynasty)." *Zhongyuan Wenwu* (3): 12–16. (in Chinese)

— 2009. "Xinzhai juluo kaogu de shijian yu fangfa (Settlement archaeology at the Xinzhai site)." *Kaogu* (2): 45–54. (in Chinese)

Zhao, C., P. Lv, J. Yuan and Y. Fang. 2012. "Wadian yizhi chutu dongwu yicun de yuansu he sitongweisu bizhi fenxi (The analysis of Sr isotope results from animal remains at Wadian)." *Kaogu* (11): 89–96. (in Chinese)

Zhao, C. and S. Zhang. 2010. "Xinzhai juluo kaogu de hugu yu zhanwang (Settlement archaeology at Xinzhai: a review and an outlook)." *Zhongyuan Wenwu* (2): 7–13. (in Chinese)

Zhao, Z. 2009. "Eastward spread of wheat into China– New data and new issues." *Chinese Archaeology* 9 (1): 1–9.

— 2011. "New archaeobotanic data for the study of the origins of agriculture in China." *Current Anthropology* 52: 295–306

Zhao, Z. 1981. "Henan Mixian Xinzhai yizhi de shijue (Trial excavation at the Xinzhai site)." *Kaogu* (5): 398–408. (in Chinese)

— 1985. "Luelun Xinzhai qi Erlitou wenhua (On the Xinzhai phase of the Erlitou culture)." In *Zhongguo Kaogu Xue Hui Di 4 Ci NianHui Lunwenji (The 4th Annual Meeting of the Chinese Archaeology*

Association): 13–18. Beijing: Wenwu Press. (in Chinese)

Zhou, B. 1984. "Zhongguo xinshiqi shidai de jiaxu (Domestic animals during the Neolithic China)." In *Xin Zhongguo de Kaogu Faxian he Yanjiu (Archaeological discoveries and studies of the PR. China)*, edited by Insititute of Archaeology CASS, 196–210. Beijing: Wenwu Press. (in Chinese)

Zhu, N. 2008. "Hongshan wenhua shoumian quexing yushi yanjiu (Study on the C-shaped beast jade from Hongshan Culture)." *Kaogu Xuebao* (1): 2–4. (in Chinese)

Zhu, Y. 2005. "Guanyu Shang dai zhongyuan diqu yesheng dongwu zhu wenti de kaocha (Issues about wild animals in the Central Plain during the Shang dynasty)." *Yindu Xuekan* (3): 1–9. (in Chinese)

Zvelebli, M. 1992. "Hunting in farming societies: the prehistoric perspective." *Anthropozoologica 16: 7-18.*

www.ingramcontent.com/pod-product-compliance
Lightning Source LLC
LaVergne TN
LVHW071218210926
848482LV00015B/842